5G Explained

5G Explained

Security and Deployment of Advanced Mobile Communications

Jyrki T. J. Penttinen
Atlanta, Georgia, USA

Registered Offices
John Wiley & Sons, Inc., 111 River Street, Hoboken, NJ 07030, USA
John Wiley & Sons Ltd, The Atrium, Southern Gate, Chichester, West Sussex, PO19 8SQ, UK

Editorial Office
The Atrium, Southern Gate, Chichester, West Sussex, PO19 8SQ, UK
For details of our global editorial offices, customer services, and more information about Wiley products visit us at www.wiley.com.

Wiley also publishes its books in a variety of electronic formats and by print-on-demand. Some content that appears in standard print versions of this book may not be available in other formats.

Library of Congress Cataloging-in-Publication Data

Names: Penttinen, Jyrki T. J., author.
Title: 5G explained : security and deployment of advanced mobile
    communications / Jyrki T. J. Penttinen, Atlanta, Georgia, USA.
Other titles: Five G explained
Description: Hoboken, NJ, USA : Wiley, [2019] | Includes bibliographical
    references and index. |
Identifiers: LCCN 2018050276 (print) | LCCN 2018052072 (ebook) | ISBN
    9781119275701 (AdobePDF) | ISBN 9781119275732 (ePub) | ISBN 9781119275688
    (hardcover)
Subjects: LCSH: Global system for mobile communications–Security measures. |
    Global system for mobile communications–Technological innovation.
Classification: LCC TK5103.483 (ebook) | LCC TK5103.483 .P46 2019 (print) |
    DDC 621.3845/6–dc23
LC record available at https://lccn.loc.gov/2018050276

Cover Design: Wiley
Cover Image: © nicomenijes/iStock.com

Set in 10/12pt WarnockPro by SPi Global, Chennai, India

Printed in the UK

Contents

Author Biography

Dr. **Jyrki T. J. Penttinen,** the author of *5G Explained*, started his activities in mobile communications industry in 1987 by evaluating early-stage NMT-900, DECT, and GSM radio network performance. After he obtained the MSc (EE) grade from the Helsinki University of Technology (HUT) in 1994, he worked for Telecom Finland (Sonera and TeliaSonera Finland) and Xfera Spain (Yoigo), performing technical tasks related to 2G and 3G. He also established and managed Finesstel Ltd. in 2002–2003, carrying out multiple consultancy and training projects in Europe and in the Americas. Afterward, he worked for Nokia and Nokia Siemens Networks in Mexico, Spain, and in the United States from 2004 to 2013. During this time with mobile network operators and equipment manufacturers, Dr. Penttinen was involved in a wide range of operational and research activities related to system and architectural design, investigation, standardization, training, and technical management. His focus and special interest were in the radio interface of cellular networks and mobile TV such as GSM, GPRS/EDGE, UMTS/HSPA, and DVB-H. From 2014 to 2018, in his position as Program Manager with G+D Mobile Security Americas, USA, his focus areas included mobile and IoT security and innovation with a special emphasis on 5G. Since 2018, he has worked for GSMA North America as Technology Manager assisting operator members with the adoption, design, development, and deployment of GSMA specifications and programmes.

Dr. Penttinen obtained his LicSc (Tech) and DSc (Tech) degrees from HUT (currently known as Aalto University, School of Science and Technology) in 1999 and 2011, respectively. In addition to his main work, he has been an active lecturer and has written dozens of technical articles and authored telecommunications books, from which those published by Wiley are *The Wireless Communications Security* (2017), *The LTE-Advanced Deployment Handbook* (2016), *The Telecommunications Handbook* (2015), *The LTE/SAE Deployment Handbook* (2011), and *The DVB-H Handbook* (2009). More information on his publications can be found at www.finesstel.com.

Preface

Mobile communication systems have vastly evolved since the introduction of the first, analogue generation in 1980s. Ever since, each new commercial system has offered novelty functions and features outperforming the older ones. The current generations have indeed been operational in a parallel fashion all this time except for the very first generation, which was decommissioned practically everywhere as we entered new millennia.

The fourth generation – the 3GPP's LTE-Advanced being the flagship of this era – has claimed its position as the most established system in global scale during 2010s. We are now looking forward to using the next, completely new generation, as has been the "tradition" for the past few decades. Based on these quite systematic cycles, it is easy to guess that the fifth generation will be again superior compared to any of its predecessors in terms of spectral efficiency, data rates, and capacity, among other important aspects essential for fluent user experiences. This time, 5G is an optimized enabler for time- and delay-sensitive applications such as virtual-reality and augmented-reality solutions.

Not only does 5G have considerable enhancements in terms of the latency and data rates, but it also takes care of a huge amount of Internet of Things (IoT) devices. It is assumed that machine-to-machine type of communications (MTC) will grow significantly during and after the first half of the 2010s. 5G is optimized by default for supporting such a big number of simultaneously communicating devices.

5G will also change the fundamental philosophy of the networks by modernizing the old reference architecture model infrastructure to support service-based architecture and virtualized environment where only the essential network functions are utilized as instances per need. For this purpose, 5G relies considerably on the increasing number of data centers on the field. They have common, virtualized hardware that paves the way for optimized utilization of physical resources while the software-based network functions can be utilized much more dynamically, efficiently, and faster compared to older network architectures that are based on dedicated hardware and software per each network element. This modernization of the core networks will provide highly useful techniques such as network slicing, which facilitates the network optimization for different use cases in a highly dynamic manner.

Historically, the mobile communications landscape has been rather fragmented, with multiple commercial systems forming each generation. At present, the telecom industry seems to be interested in a much more unified mobile communication system, which indeed can be achieved by the deployment of 3GPP-defined 5G networks in a global scale. We might thus finally see a truly unique and single standard defining the new generation, which will ease the interoperability and is also beneficial for customers and multiple stakeholders thanks to the expected economies of scale.

Academia has contributed strongly to the investigation of novelty candidate technologies for 5G radio and core networks while the industry has developed and tested shortlisting the most feasible concepts. Technical performance of these candidates has been under thorough testing during the pilots and trials, while the technical 3GPP specifications defining 5G system have been maturing. This high level of industry interest has been beneficial for the standardization to maintain and even expedite the original development schedules. As a result, Release 15 of 3GPP was frozen in June 2018, and after final adjustments, it is ready for providing truly interoperable solutions to equipment manufacturers and mobile network operators.

It should be noted that the 3GPP Release 15 represents merely the first phase of 5G, which works for the introduction of key 5G services while 3GPP still maintains and enhances technical specifications of the parallel systems for 2G (GSM), 3G (UMTS/HSPA), and 4G (LTE/LTE-Advanced).

There will be new and enhanced 5G specifications, too, to comply with the strict 5G functional and performance requirements of the IMT-2020 (International Mobile Telecommunications for 5G), which has been defined by ITU (International Telecommunications Union). After the late drops referring to the additional items in Release 15, the second phase of 3GPP's 5G will be published in the technical specifications of Release 16 by the beginning of the next decade. So, the first 5G networks complying fully with the IMT-2020 requirements are expected to be deployed as of 2020.

The 3GPP Release 16 will bring along many new functionalities, enhancing the initial 5G performance. This phased approach provides means for a fluent and expedited deployment of 5G services based on the previous core infrastructure in the initial deployments, as defined in the 4G Enhanced Packet Core (EPC) specifications of 3GPP. One example of this hybrid mode is the set of non-standalone modes, which the mobile network operators can deploy selectively while waiting for the 3GPP Release 16-based solutions.

While the 5G specifications have been under development through the second half of 2018, there has already been plenty of speculative information available on 5G in printed form. Now, as the 3GPP Release 15 has been frozen, this book summarizes concretely the essential aspects of 5G based on the latest knowledge interpreted from the specifications and industry. This book presents the overall concept of 5G, helping the reader to understand the "big picture" of the theme and presenting focused points on security and deployment aspects.

I hope you enjoy the contents of this book in your preparation for the exiting journey in exploring yet another mobile generation! As has been the case with my previous books published by Wiley, I would highly appreciate all your feedback via my email address, jyrki.penttinen@hotmail.com.

Jyrki T. J. Penttinen
Atlanta, GA, USA

Acknowledgments

This *5G Explained* complements my previous five books published with Wiley on telecommunication technologies since 2009. Looking back at the development, it is fascinating to realize how the systems evolve with such an overwhelming pace, providing us users with better performance and new, more interesting functionalities. As we approach the 5G era, this is especially clear with such groundbreaking, new principles applied in the networks.

Along with this sixth book, I express my warmest thanks for all the support to the Wiley teams I have worked with throughout the respective 10-year period. As for specifically this *5G Explained* book, I want to give special thanks to Ms. Sandra Grayson for such great support and editorial guidance. I thank also Ms. Cheryl Ferguson for the editing, Ms. Sonali M. Melwani for all the coordination in shaping the manuscript into the final book, and Ms. Nithya Sechin and Apoorva Sindoori for keeping track of the advances.

One quite important part of the security section of this book would not have been possible to summarize without the support of my colleagues of Giesecke+Devrient at G+D Mobile Security. I want to express my special thanks to Mr. Claus Dietze, who contributed an important base to the security chapter. Knowing this list will not even come close to being complete, I also want to extend my gratitude to my former colleagues of Giesecke+Devrient for all the support specifically related to 5G, which eased my way in understanding and documenting aspects that I believe will be of utmost importance in the 5G era. Please note, though, that this book has been accomplished by myself in my personal capacity as an author. The opinions expressed in this book are thus my own and do not necessarily reflect the view of my current or past employers.

As has already been kind of "tradition," I have done this work during my spare time. I am thus thankful for all the support and patience of my close family, Paloma, Katriina, Pertti, Stephanie, Carolyne, and Miguel, and all the ones on my side who encouraged me to continue to pursue this passion.

Jyrki T. J. Penttinen
Atlanta, GA, USA

Abbreviation List

1G	first generation of mobile communications; analogue systems
2G	second generation of mobile communications; digital systems
3G	third generation of mobile communications; multimedia-capable systems
3GPP	3rd Generation Partnership Project
4G	fourth generation of mobile communications; enhanced multimedia systems
5G	fifth generation of mobile communications; systems suitable for connected society
5GAA	5G Automotive Association
5G-EIR	5G Equipment Identity Register
5G-GUTI	5G Globally Unique Temporary Identity
MM sub	Mobility Management sublayer
5G-PPP	5G Infrastructure Public-Private Partnership
5GS	5G system
6LoWPAN	IPv6 Low-power wireless personal area network
AAA	authentication, authorization, and accounting
AAS	active antenna system
ADAS	advanced driver assistance system
AERIS	Applications for the Environment: Real-Time Information Synthesis
AES	Advanced Encryption Standard
AF	application function
AI	artificial intelligence
AID	application identifier
AKA	Authentication and Key Agreement
AM	Acknowledge Mode
AMBR	Aggregate Maximum Bit Rate
AMF	Access and Mobility Management Function
AMPS	Advanced Mobile Phone System
AN	access network
ANR	Automatic Neighbor Relations
AP	access point
AP	Application Protocol
API	application programming interface
ARIB	Association of Radio Industries and Businesses in Japan

ARP	auto radio phone
ARPF	Authentication Credential Repository and Processing Function
AS	access stratum
ASA	authorized shared access
ASN.1	Abstract Syntax Notation One
ATIS	Alliance for Telecommunications Industry Solutions in the U.S.
AU	application unit
AUSF	Authentication Server Function
AV	authentication vectors
AWGN	Additive White Gaussian Noise
BDS	BeiDou Navigation Satellite System
BH	backhaul
BPSK	Binary Phase Shift Keying
BS	base station
BSC	base station controller
BSM	basic safety messages
BTS	base transceiver station
C2C CC	Car-to-Car Communication Consortium
CA	carrier aggregation
CA	certification authority
CAD	connected and automated driving
CAPIF	Common API Framework
Cat	category (IoT)
CBRS	Citizens Broadband Radio Service
CC	common criteria (ISO/IEC)
CCSA	China Communications Standards Association
CDF	cumulative distribution function
CDR	charging data record
CEPT	Conférence Européenne des Postes et des Télécommunications
CID	cell ID
C-IoT	cellular IoT
C-ITS	Cooperative Intelligent Transport System
CIP	Critical Infrastructure Protection
CK	ciphering key
CM	connection management
CM sub	connection management sublayer
CN	core network
CNT	computer network technologies
CoAP	Constrained Application Protocol
CoMP	coordinated multipoint
CORD	Central Office Re-architected as Data Center
COTS	commercial off-the-shelf
CP	control plane
CP	cyclic prefix
CPA	certified public accountants
CPS	cyber physical security
CPU	central processing unit

CRC	Cyclic Redundancy Check
CriC	critical communications
CSA	Cloud Security Alliance
CSC	Cloud Service Category
CSI-RS	channel-state information reference signal
CSMA-CA	Carrier Sense Multiple Access and Collision Avoidance
CSN	Connectivity Service Network
CSP	Cloud Service Providers
CTL	control
CU	centralized unit
CUPS	Control and User Plane Separation (EPC nodes)
C-V2X	cellular V2X
DC	dual connectivity
DCI	downlink control information
DDoS	distributed denial of service
DECT	Digital Enhanced Cordless Telecommunications
DHCP	Dynamic Host Control Protocol
DL	downlink
DM	device management
DM-RS	demodulation reference signals
DN	data network
DNN	Data Network Name
DNS	Dynamic Name Server
DoD	Department of Defense (USA)
DPI	deep packet inspection
DRB	data radio bearer
DRM	digital rights management
DSRC	dedicated short-range communications
DSS	Data Security Standard
DTF	discrete Fourier transform
DU	distributed unit
E CID	enhanced cell ID
EAP	Extensible Authentication Protocol
EC	European Commission
ECC	Electronic Communications Committee
EC-GSM	enhanced coverage GSM (IoT)
ECM	EPS Connection Management
ECO	European Communications Office
EDGE	Enhanced Data rates for Global evolution
EE	energy efficiency
EGPRS	enhanced GPRS
EIRP	Effective Isotropic Radiated Power
eLAA	enhanced LAA
eLTE	Evolved LTE
eMBB	evolved Multimedia Broadband
eMBMS	evolved MBMS
EMM	EPS Mobility Management

eMTC	evolved Machine-Type Communication
eNodeB	evolved NodeB (eNB)
EPC	Evolved Packet Core (4G)
EPS	Enhanced Packet System
eSIM	embedded subscriber identity module
E-SMLC	Evolved Serving Mobile Location Centre
ETN	Edge Transport Node
ETSI	European Telecommunications Standards Institute
EU	European Union
eUICC	embedded UICC
EUM	eUICC Manufacturer
E-UTRA	Evolved UTRA
E-UTRAN	Evolved UMTS Terrestrial Radio Access Network
FBMC	Filter Bank Multicarrier
FCC	Federal Communications Commission (USA)
FDD	frequency division duplex
FDM	frequency division multiplexing
FEC	Forward Error Correction
FF	form factor
FFT	fast Fourier transform
FG	Forwarding Graph (NF)
FH	fronthaul
FIPS	Federal Information Processing Standards
FM	fault management
FMVSS	Federal Motor Vehicle Safety Standard
FR	frequency range
GAA	General Authorized Access (CBRS)
GBR	guaranteed bit rate
GGSN	Gateway GPRS Support Node
GMLC	Gateway Mobile Location Center
gNB	5G NodeB
GNSS	Global Navigation Satellites System
GP	Global Platform
GPRS	General Packet Radio Service
GPS	global positioning system
GSC	Global Standards Collaboration
GSM	Global System for Mobile communications
GSMA	GSM Association
gsmSCF	GSM Service Control Function
GTP	GPRS Tunneling Protocol
GTP U	GPRS Tunneling Protocol in user plane
GWCN	gateway core network
HARQ	hybrid automatic repeat and request
HCE	host card emulation
HE AV	home environment AV
HeNodeB	home eNodeB
HG	home gateway

HIPAA	Health Insurance Portability and Accountability Act
HLS	higher layer split (gNB)
hNRF	NRF in the home PLMN
HPLMN	home public land mobile network
HR	home routed
HS	hot spot
hSEPP	Home Security Edge Protection Proxy
HSM	hardware security module
HSPA	High Speed Packet Access
HSS	Home Subscription Server
H-SMF	home SMF
HW	hardware
I²C	Inter-Integrated Circuit
IaaS	Infrastructure as a Service
IATN	Inter-Area Transport Node
ICC	Integrated Circuit Cards
ICI	Inter-Carrier Interference
ICT	information and communication technology
IDFT	inverse discrete Fourier transform
IE	Information Element
IEC	International Electrotechnical Commission
IEEE	Institute of Electrical and Electronics Engineers
IEEE-SA	IEEE Standards Association
IETF	Internet Engineering Task Force
IFFT	inverse fast Fourier transform
IK	integrity key
IMS	IP Multimedia Subsystem
IMT-2000	International Mobile Telecommunications (3G)
IoT	Internet of Things
IoT-GSI	Global Standards Initiative on Internet of Things
IP	Internet Protocol
IPX	Internet Protocol Packet eXchange
ISA	International Society for Automation
ISG	Industry Specification Group
ISI	inter-symbol interference
ISO	International Standardization Organisation
ISP	Internet service provider
iSSP	integrated SSP
IT	information technology
ITL	image trusted loader
ITS	Intelligent Transportations Systems
ITU	International Telecommunication Union
ITU-R	radio section of ITU
ITU-T	telecommunication sector of ITU
iUICC	integrated UICC
IWMSC	inter-working MSC
JTC	Joint Technical Committee (ISO/IEC)

KDF	Key Derivation Function
K*n*	key, *n* refers to key's purpose, see Chapter 8, Table 3
KPI	key performance indicator
LAA	License Assisted Access
LBO	local breakout
LBS	location-based service
LDPC	Low-Density Parity Check
LI	Lawful Interception
LLS	lower-layer split (gNB)
LOS	line of sight
LPWAN	low-power wide area network
LSA	Licensed Shared Access
LTE	long-term evolution
LTE-A	LTE-Advanced
LTE-M	IoT-mode of LTE
LTE-U	unlicensed LTE band
M2M	machine-to-machine
MAC	Medium Access Control
MANET	Mobile Ad-hoc Network
MAP	Mobile Application Part
MBB	Mobile Broadband
MBMS	Multimedia Broadcast Multicast Service
MBS	Metropolitan Beacon System
MCC	Mobile Country Code
MCE	Mobile Cloud Engine
MCPTT	mission-critical push-to-talk
MCS	Modulation and Coding Scheme
MDT	mobile data terminal
ME	mobile equipment
MEC	Mobile Edge Computing
METIS	Mobile and wireless communications Enablers for Twenty-twenty (2020) Information Society
MIMO	multiple-in, multiple-out
mIoT	massive Internet of Things
MITM	man in the middle
MM sub	Mobility Management sublayer
MM	Mobility Management
MME	Mobility Management Entity
mMTC	massive machine-type communications
MN	master node
MNC	Mobile Network Code
MNO	mobile network operator
MO	mobile originated (SMS)
MOCN	Multi Operator Core Network
MORAN	Multiple Operator RAN
MR-DC	Multi-RAT Dual Connectivity
MS	mobile station

MSC	Mobile Services Switching Center
MSIN	Mobile Subscription Identification Number
MT	mobile terminated (SMS)
MTA	Mobile Telephony System (version A)
MTC	machine-type communications
MU-MIMO	Multi User MIMO
MUX	multiplexer
MVNO	Mobile Virtual Network Operator
N3IWF	Non-3GPP Interworking Function
NaaS	Network as a Service
NAS	non-access stratum
NB-IoT	narrow-band IoT
NCR	neighbor cell relation
NDS	Network Domain Security
NEA	NR encryption algorithm (NEA0…3)
NEF	Network Exposure Function
NEO	network operations
NERC	North American Electric Reliability Corporation
NF	network function
NF	noise figure
NFC	near-field communications
NFV	network functions virtualization
NG-AP	NG Application Protocol
NG	Next Generation
NGC	Next Generation Core (5G)
NGCN	Next Generation Core Network
ng-eNB	Next Generation evolved NodeB (enhanced 4G eNB)
NGMN	Next Generation Mobile Networks
NG-RAN	Next Generation Radio Access Network
NH	next hop
NHTSA	National Highway Transportation and Safety Administration
NIA	NG integrity algorithm (NIA0…3)
NIST	National Institute of Standards and Technology (USA)
NLOS	non-line of sight
NMT	Nordic Mobile Telephony
NOMA	nonorthogonal multiple access
NPRM	notice of proposed rulemaking
NR	New Radio (5G)
NRF	Network Repository Function
NSSAI	Network Slice Selection Assistance Information
NSSF	Network Slice Selection Function
NVM	non-volatile memory
NWDA	Network Data Analytics
NWDAF	Network Data Analytics Function
O&A	operations and maintenance
OBU	on-board unit
OCP	Open Compute Project

OEM	original equipment manufacturer
OFDM	orthogonal frequency division multiplexing
OFDMA	OFDM Access
OFL	Open Firmware Loader
OLA	operating level agreement
OMA	Open Mobile Alliance
ONAP	Open Network Automation Platform
OOB	out-of-band leakage
OS	operating system
OSI	Open System Interconnect
OSS	operations support system
OTA	over the air
OTDOA	observed time difference of arrival
OTP	one-time programmable
OWASP	Open Web Application Security Project
PaaS	Platform-as-a-Service
PAL	priority access license (CBRS)
PAPR	peak-to-average power ratio
PBCH	Physical Broadcast Channel
PCF	Policy Control Function
PCI	payment card industry
PCRF	Policy and Charging Enforcement Function
PDCCH	Physical Downlink Control Channel
PDCP	Packet Data Convergence Protocol
PDN	Packet Data Network
PDSCH	Physical Downlink Shared Channel
PDU	packet data unit
PEI	Permanent Equipment Identity
PFCP	packet forwarding control plane
PFDF	Packet Flow Descriptions Function
P-GW	Proxy Gateway
PHI	protected health information
PKI	public key infrastructure
PLMN	public land mobile network
PM	performance monitoring
PNF	Physical Network Functions
PoC	proof of concept
POS	point of sales
PRACH	Physical Random-Access Channel
PRS	positioning reference signals
PSK	phase shift keying
PSM	power save mode
PSS	primary synchronization signal
PTP	point-to-point
PT-RS	Phase-Tracking Reference Signals
PUCCH	Physical Uplink Control Channel
PUSCH	Physical Uplink Shared Channel

PWS	Public Warning System
QAM	quadrature amplitude modulation
QCI	QoS class identifier
QoE	quality of experience
QoS	quality of service
QPSK	Quadrature Phase Shift Keying
QZSS	Quasi Zenith Satellite System
RAM	random access memory
RAN	radio access network
RAT	radio access technology
RF	radio frequency
RLB	radio link budget
RLC	Radio Link Control
RM	registration management
RN	remote node
RNC	Radio Network Controller
RNL	Radio Network Layer
ROM	read-only memory
RPMA	Random Phase Multiple Access
RPO	Recovery Point Objective
RR sub	radio resource management sublayer
RRC	Radio Resource Control
RRH	remote radio head
RRM	Radio Resource Management
RSP	remote SIM provisioning
RSU	roadside unit
RT	ray tracing
RTO	recovery time objective
SA	Secure Appliance
SA	security association
SA	standalone
SA	system architecture
SAE	System Architecture Evolution
SAP	service access point
SAS	Security Accreditation Scheme (GSMA)
SAS-SM	SAS for subscription management
SAS-UP	SAS for UICC production
SBA	service-based architecture
SBAS	space-based augmentation systems
SBC	session border controller
SC&C	Smart Cities and Communities (ITU)
SCA	Smart Card Alliance (currently STA)
SCEF	Service Capability Exposure Function
SC-FDM	single-carrier frequency division multiplex
SC-FDMA	single-carrier frequency division multiple access
SCG	Secondary Cell Group
SCM	security context management

SCMF	Security Context Management Function
SCP	smart card platform (ETSI)
SCTP	Stream Control Transmission Protocol
SDCI	Sponsored Data Connectivity Improvements
SDF	Service Data Flow
SDN	software-defined networking
SDO	standards developing organizations
SDS	structured data storage
SDU	Service Data Unit
SE	secure element
SEAF	Security Anchor Function
SEL	spectral efficiency loss
SEPP	Security Edge Protection Proxy
SGSN	Serving GPRS Support Node
S-GW	Serving Gateway
SIDF	Subscription Identifier De-Concealing Function
SIM	subscriber identity module
SiP	system in package
SISO	single-input, single-output
SLA	service-level agreement
SLR	service-level reporting
SM AL	Short Message Application Layer
SM CP	Short Message Control Protocol
SMI	Structure of Management Information
SM RL	Short Message Relay Layer
SM RP	Short Message Relay Protocol
SM TL	Short Message Transfer Layer
SM	session management
SMC	Short Message Control
SMF	Session Management Function
SMR	Short Message Relay
SMS	Short Message Service
SMSF	Short Message Service Function
SN	secondary node; slave node
SNSSAI	Single Network Slice Selection Assistance Information
SOC	service organization control
SoC	system on chip
SON	self-optimizing network
SPCF	Security Policy Control Function
SPI	Serial Peripheral Interface
SQL	Structured Query Language
SRS	Sounding Reference Signal
SSAE16	Statement on Standards for Attestation Engagements 16
SSC	Session and Service Continuity (mode)
SSP	Secure Smart Platform
SSS	secondary synchronization signal
STA	Secure Technology Alliance

STAR	Security, Trust & Assurance Registry (CSA)
SUCI	Subscriber Concealed Identifier
SUL	supplementary uplink
SU-MIMO	single-user MIMO
SUPI	Subscription Permanent Identifier
SUPL	secure user plane location
SW	software
TA	tracking area
TACS	Total Access Communications System
TB	transport block
TBS	Terrestrial Beacon System
TC	Technical Committee (of ETSI)
TCAP	Transaction Capabilities Application Part
TCG	Trusted Computing Group
TCO	total cost of ownership
TDD	Time Division Duplex
TDF	Traffic Detection Function
TDM	Time Division Multiplexing
TDMA	Time Division Multiple Access
TEE	Trusted Execution Environment
TIA	Telecommunications Industry Association (USA)
TIF	Transport Intelligent Function
TIP	Telecom Infra Project
TI-SCCP	Transport Independent Signaling Connection Control Part
TLS	Transport Layer Security
TLV	type-length-value
TN	transport node
TNL	transport network layer
TP	transmission point
TPM	Trusted Platform Module
TRE	tamper-resistant element
TRxP	Transmission Reception Point
TSAG	Telecommunication Standardization Advisory Group
TSDSI	Telecommunications Standards Development Society in India
TSG	Technical Specification Group (3GPP)
TSON	Time-Shared Optical Network
TTA	Telecommunications Technology Association in Korea
TTC	Telecommunication Technology Committee in Japan
UDC	Uplink Data Compression
UDM	unified data management
UDP	User Datagram Protocol
UDR	Unified Data Repository
UDSF	Unstructured Data Storage Function
UE	user equipment
UFMC	Universal Filter Multi-Carrier
UHF	ultra-high frequency
UICC	universal integrated circuit card

UL	uplink
UL-CL	uplink classifier
UM	Unacknowledged Mode
UMTS	Universal Mobile Telecommunications System
UP	user plane
UPF	User Plane Function
UPS	unbreakable power system
URLLC	ultra-reliable low latency communications
USAT	USIM Application Toolkit
USB	universal serial bus
USIM	universal SIM
UTRAN	UMTS Terrestrial Radio Access Network
UX	user experience
V2I	vehicle-to-infrastructure
V2P	vehicle-to-pedestrian
V2R	vehicle-to-roadside
V2V	vehicle-to-vehicle
V2X	vehicle-to-everything
V5GTF	Verizon 5G Technology Forum
VANET	Vehicular Ad-hoc Network (car-to-car communications)
vCDN	virtual Content Delivery Network
vEPC	virtual evolved packet core
vIMS	virtual IP Multimedia Subsystem
VM	virtual machine
VNF	Virtual Network Functions
vNRF	NRF in the visited PLMN
VPLMN	visited PLMN
VPN	virtual private network
VPP	Virtual Primary Platform
VR	virtual reality
vSEPP	visited network's security proxy
V-SMF	visited SMF
vUICC	virtual UICC
WAVE	Wireless Access in Vehicular Environments
WCDMA	Wideband Code Division Multiple Access
WDM-PON	Wavelength Division Multiplexing Passive Optical Network
WGFM	Working Group for Frequency Management (ECC)
WiMAX	Worldwide Interoperability for Microwave Access
WISP	wireless Internet service provider
WLAN	Wireless Local Area Network
WRC	World Radio Conference

1

Introduction

1.1   Overview

The 5G Explained presents key aspects of the next, evolved mobile communications system after the 4G era. This book concentrates on the deployment of 5G and discusses the security-related aspects whilst concrete guidelines of both topics for the earlier generations can be found in the previously published books of the author in Refs. [1, 2].

The fifth generation is a result of long development of mobile communications, the roots of its predecessors dating back to the 1980s when the first-generation mobile communication networks started to convert into a reality [3]. Ever since, the new generations up to 4G have been based on the earlier experiences and learnings, giving the developers a base for designing enhanced security and technologies for the access, transport, signaling, and overall performance of the systems.

Regardless of the high performance of 4G systems, the telecom industry has identified a need for faster end-user data rates due to constantly increasing performance requirements of the evolving multimedia. 5G systems have thus been designed to cope with these challenges by providing more capacity and enhanced user experiences that solve all the current needs even for the most advanced virtual reality applications. At the same time, the exponentially enhancing and growing number of IoT (Internet of Things) devices requires new security measures such as security breach monitoring, prevention mechanisms, and novelty manners to tackle the vast challenges the current and forthcoming IoT devices bring along.

The demand for 5G is reality based on the major operators' interest to proof the related concepts in global level. Nevertheless, the complete variant of 5G is still under development, with expected deployments complying with the full set of the strict performance requirements taking place as of 2020.

As there have been more concrete development and field testing activities by major operators, as well as agreements for the forthcoming 5G frequency allocation regulation by International Telecommunications Union (ITU) World Radio Conference (WRC) 19, this book aims to summarize recent advances in the practical and standardization fields for detailing the technical functionality, including the less commonly discussed security-breach prevention, network planning, optimization, and deployment aspects of 5G based on the available information during 2018 and basing on the first phase of the 3rd Generation Partnership Project (3GPP) Release 15, which is the starting point for the gradual 5G deployment.

5G Explained: Security and Deployment of Advanced Mobile Communications, First Edition.
Jyrki T. J. Penttinen.
© 2019 John Wiley & Sons Ltd. Published 2019 by John Wiley & Sons Ltd.

1.2 What Is 5G?

The term 5G refers to the fifth generation of mobile communication systems. They belong to the next major phase of mobile telecommunications standards beyond the current 4G networks that will comply with the forthcoming International Mobile Telecommunications (IMT)-2020 requirements of ITU-R (radio section of the International Telecommunications Union). 5G provides much faster data rates with very low latency compared to the current systems up to 4G. It thus facilitates the adaptation of highly advanced services in wireless environment.

The industry seems to agree that 5G is, in fact, a combination of novel (yet to be developed and standardized) solutions and existing systems basing on 4G Long-Term Evolution (LTE)-Advanced, as well as non-3GPP access technologies such as Wi-Fi, which jointly contributes to optimizing the performance (providing at least 10 times higher data rate compared to current LTE-Advanced networks), lower latency (including single-digit range in terms of millisecond), and support of increased capacity demands for huge amounts of simultaneously connected consumer and machine-to-machine, or M2M, devices. Because of the key enablers of 5G, some of the expected highly enhanced use cases would include also the support of tactile Internet and augmented, virtual reality, which provide completely new, fluent, and highly attractive user experiences never seen before.

At present, there are many ideas about the more concrete form of 5G. Various mobile network operators (MNOs) and device manufacturers have been driving the technology via concrete demos and trials, which has been beneficial for the selection of optimal solutions in standardization. This, in turn, has expedited the system definition schedules. While these activities were beneficial for the overall development of 5G, they represented proprietary solutions until the international standardization has ensured the jointly agreed 5G definitions, which, in turn, has led into global 5G interoperability.

The mobile communication systems have converted our lives in such a dramatic way that it is hard to imagine communication in the 1980s, when facsimiles, letters, and plain old fixed-line telephones were the means for exchanging messages. As soon as the first-generation mobile networks took off and the second generation proved the benefits of data communications, there was no returning to those historical days. The multimedia-capable third generation in the 2000s, and the current, highly advanced fourth generation offer us more fluent always-on experiences, amazing data rates, and completely new and innovative mobile services. The pace has been breathtaking, yet we still are in rather basic phase compared to the advances we'll see during the next decade. We are in fact witnessing groundbreaking transition from the digital world toward truly connected society that will provide us with totally new ways to experience virtual reality and ambient intelligence of the autonomic IoT communications.

The ongoing work on the development of the next big step in the mobile communications, the fifth generation, includes the IoT as an integral part. Although one of the key goals of the 5G is to provide considerably higher data rates compared to the current 4G systems, with close to zero delays, at least an equally important aspect of the new system will be the ability to manage huge amount of simultaneously communicating IoT devices – perhaps thousands under a single radio cell.

1.3   Background

The term *5G* is confusing. During 2016–2017, there were countless public announcements on the expected 5G network deployments while the 4G deployment was still in its most active deployment phase. Up to the third-generation mobile communication networks, the terminology has been quite understandable, as 3G refers to a set of systems that comply with the IMT-2000 (International Mobile Telecommunications for 3G) requirements designed by the ITU. Thus, the cdma2000, Universal Mobile Telecommunications System (UMTS)/High Speed Packet Access (HSPA) and their respective evolved systems belong to the third generation as the main representatives of this era.

The definition of the fourth generation is equally straightforward, based on the ITU's IMT-Advanced requirements. While 3G had multiple representatives in practice, there are only two systems fulfilling the official, globally recognized 4G category as defined in IMT-Advanced, and they are the 3GPP LTE-Advanced as of Release 10, and the IEEE 802.16m referred to also as WiMAX2. The first 3GPP Release 8 and Release 9 LTE networks were deployed in 2010–2011, and their most active commercialization phase took place around 2012–2014. Referring to ITU-terminology, these networks prior to Release 10 still represented the evolved 3G era, which, as soon as they were upgraded, resulted in the fully compatible 4G systems.

While 4G was still being developed, the 5G era generated big interest. The year 2017 was a concrete show-time for many companies for demonstrating how far the technical limits could be pushed. Some examples of these initiations, among many others, included Verizon 5G Technology Forum, which included partners in the Verizon innovation centers [4], and Qualcomm, which demonstrated the capabilities of LTE-Advanced Pro via millimeter-wave setup [5].

These examples and other demos and field trials prior to the commercial deployment of 5G indicated the considerably enhanced performance and capacity that the 5G provides, although fully deployed, Phase 2 of 5G as defined by 3GPP is still set to the 2020 time frame. As soon as available, the 5G era will represent something much more than merely a set of high-performance mobile networks. It will, in fact, pave the way for enabling a seamlessly connected society with important capabilities to connect a large number of always-on IoT devices.

The idea of 5G is to rely on both old and new technologies on licensed and unlicensed radio frequency (RF) bands that are extended up to several GHz bands to bring together people, things, data, apps, transport systems and complete cities, to mention only some – in other words, everything that can be connected. The 5G thus functions as a platform for ensuring smooth development of the IoT, and it also acts as an enabler for smart networked communications. This is one of the key statements of ITU, which eases this development via the IMT-2020 vision.

The important goal of 5G standard is to provide interoperability between networks and devices, to offer high capacity energy-efficient and secure systems, and to remarkably increase the data rates with much less delay in the response time. Nevertheless, the 5th generation still represents a set of ideas for highly evolved system beyond the 4G. As has been the case with the previous generations, the ITU has taken an active role in coordinating the global development of the 5G.

1.4 Research

There are many ideas about the form of 5G. Major operators and device manufacturers have actively conducted technology investigations, demos, and trials aiming to prove the concepts and contributing to the standardization.

There are also several research programs established to study the feasibility and performance of new ideas in academic level. As an example, the European Union (EU) coordinates 5G research programs under various teams. More information about the latest European Commission (EC) funded 5G research plans can be found in EU web page, which summarizes 5G initiatives [6]. As stated by EU, the 5G of telecommunications systems will be the most critical building block of our digital society in 2020–2030. Europe has taken significant steps to lead global developments toward this strategic technology. Furthermore, EU has recognized that the 5G will be the first instance of a truly converged network environment where wired and wireless communications will use the same infrastructure, driving the future networked society. EU states that *5G will provide virtually ubiquitous, ultra-high bandwidth connectivity not only to individual users but also to connected objects. Therefore, it is expected that the future 5G infrastructure will serve a wide range of applications and sectors, including professional uses such as assisted driving, eHealth, energy management, and possibly safety applications.*

The EC study programs include FP7 teams and METIS (Mobile and wireless communications Enablers for Twenty-twenty (2020) Information Society), and other internationally recognized entities. One of the international joint activities is the cooperation between EU and Brazil [7].

As for the 5G radio capacity needs on the current bands, the European Commission aims to coordinate the use of the 700 MHz band for mobile services to provide higher-speed and higher-quality broadband and cover wider areas, including rural and remote regions. The concrete goal of EU is to provide mobile broadband speeds beyond 100 Mb/s.

1.5 Challenges for Electronics

One of the expected key abilities of the 5G networks is the high-energy efficiency to cope with a big amount of low-power IoT devices in the field. The benefits include better cost-efficiency, sustainability, and widening the network coverage to remote areas. Some of the base technologies for facilitating the low energy include advanced beamforming as well as radio interface optimization via user-data and system-control plane separation. Other technologies include reliance on virtualized networks and clouds.

The systems also need to be developed at the component level for both networks and devices. Autonomously functioning remote IoT devices require special attention, as they must function reliably typically several years without human interaction or maintenance. The advances in the more efficient battery technologies are thus in key position. Also, the very small devices such as consumer wearables and M2M sensor equipment may require much smaller electronic component form factors, including tiny wafer-level subscriber modules that still comply with the demanding reliability and durability requirements in harsh conditions. At the same time, the need for enhanced security aspects will require innovative solutions in the hardware (HW) and software (SW) levels.

1.6 Expected 5G in Practice

The 5G is a result of a long development of mobile communications, with roots going back to the 1980s when the first-generation mobile communications networks began to be reality. Ever since the first data services, which the 2G systems started to include around the mid-1990s, the new generations up to 4G have been based on the earlier experiences and learnings, giving the developers a base for designing enhanced technologies for the access, transport, signaling, and overall performance of the systems. Figure 1.1 depicts the development of the data rates of the 3G, 4G, and 5G systems.

However, the telecom industry has identified a further need for considerably faster end-user data rates to cope with the demands of the evolving multimedia. The 5G could handle these challenging capacity requirements to provide fluent user experiences even for the most advanced virtual reality applications. At the same time, the exponentially growing number of the IoT devices require new security measures, including potential security-breach monitoring and prevention.

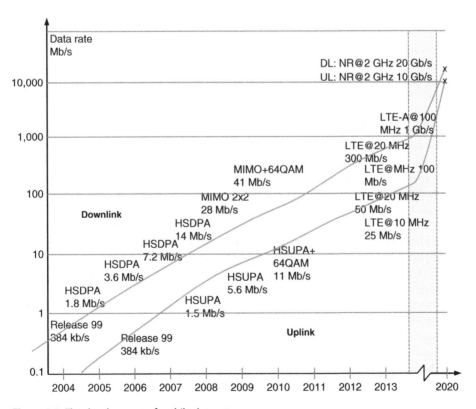

Figure 1.1 The development of mobile data rates.

Along with the new M2M and IoT applications and services, there will be role-changing technologies developed to support and complement the existing ones. The 5G is one of the most logical bases for managing this environment, together with the legacy systems in the markets.

Although the 5G is still in its infancy until the ITU officially dictates its requirements and selects the suitable technologies from the candidates, the 5G systems will be reality soon. During the deployment and operation of 5G networks, we can expect to see many novelty solutions such as highly integrated wearable devices, household appliances, industry solutions, robotics, self-driving cars, virtual reality, and other advanced, always-on technologies that benefit greatly from enabling 5G platforms.

In addition to the "traditional" type of IoT devices such as wearable devices with integrated mobile communications systems (smart watch), car communications systems and utility meters, there are also emerging technology areas such as self-driving cars that require high reliability as for the functionality as well as for the secure communications, which 5G can tackle.

As Next Generation Mobile Networks (NGMN) states in [8], the 5G will address the demands and business contexts as of 2020 by enabling a fully mobile and connected society. This facilitates the socioeconomic transformations contributing to productivity, sustainability, and overall well-being. This is achieved via a huge growth in connectivity and volume of data communications. This, in turn, is possible to provide via advanced, multilayer densification in the radio network planning and providing much faster data throughput, considerably lower latency, and higher reliability and density of simultaneously communicating devices.

In addition, to manage this new, highly complex environment, new means for managing and controlling the heterogeneous and highly energy-efficient environment is needed. One of the major needs is to ensure the proper security of the new 5G services and infrastructure, including protection of identity and privacy.

Another aspect in the advanced 5G system is the clearly better flexibility compared to any of the previous mobile communication generations. This refers to the optimal network resource utilization and providing new business models for variety of new stakeholders. This means that the 5G network functionality will be highly modular, which facilitates cost-efficient, on-demand scalability.

In practice, the above-mentioned goals are possible to achieve only via renewed radio interfaces, including totally new, higher frequencies and capacity enhancements for the accommodation of increasing the customer base in consumer markets as well as support for the expected huge amount of simultaneously communication IoT devices.

The 5G network infrastructure will be more heterogeneous than ever before, so there will be a variety of access technologies, end-user devices, and network types characterized by deeper multilayering. The challenge in this new environment is to provide to the end-users as seamless a user experience as possible.

To achieve the practical deployment schedule for the mature 5G initiating the commercial era by 2020, 3GPP as well as supporting entities such as NGMN are collaborating with the industry and relevant standardization organizations covering both "traditional" teams as well as new, open-source-based standardization bodies.

1.7   5G and Security

As for the security assurance of the new 5G era, there can be impacts expected in the "traditional" forms of SIM (Subscriber Identity Module), Universal Integrated Circuit Card (UICC), and subscription types, as the environment will be much more dynamic. The ongoing efforts in developing interoperable subscription management solutions that respond in near real time for changing devices and operators upon the need of the users are creating one of the building blocks for the always connected society. It is still to be seen what the consumer and M2M devices will look like physically in the 5G era, but we might see much more variety compared to any previous mobile network generations, including multiple wearable devices per user and highly advanced control and monitoring equipment.

Along with these completely new types of devices, the role of the removable subscription identity modules such as SIM/UICC can change; the much smaller personal devices require smaller form factors. At the same time, the techniques to tackle with the constantly changing subscriptions between devices need to be developed further, as do their security solutions. The cloud-based security such as tokenization and host card emulation (HCE), as well as the development of the device-based technologies like Trusted Execution Environment (TEE) may be in key positions in the 5G era, although the traditional SIM/UICC can still act as a base for the high security demands.

1.8   Motivations

One might wonder why yet another mobile communications generation is really needed. In fact, the fourth generation already provides quite impressive performance with low latency and high data rates.

The reasons are many-folded. Not only the increasing utilization of the mobile communications networks for ever-advancing applications including higher-definition video, virtual reality, and artificial intelligence require much more capacity even in remote areas, but – and one might argue if even primarily – the need is derived from the exponential increase of the IoT devices. The number of the intelligent sensors and other machines communicating with each other, service back-ends require support for much more simultaneously connected devices. The amount may be, as stated in one of the core requirements of the ITU, one million devices per km^2, which outnumbers clearly even the theoretically achievable capacity the advanced 4G can offer. The 5G would thus benefit especially massive IoT development, the increased data rates for consumers being another positive outcome of the new technology.

1.9   5G Standardization and Regulation

1.9.1   ITU

The ITU-R is the highest-level authority for defining the universal principles of 5G. The ITU is thus planning to produce a set of requirements for the official 5G-capable systems under the term IMT-2020. As the term indicates, the commercial systems are assumed to be ready for deployment as of 2020. This follows the logical path for ITU-defined 3G and 4G, as depicted in Figure 1.2.

The IMT-2020 is in practice a program to describe the 5G as a next-evolution step after the IMT-2000 and IMT-Advanced, and it also sets the stage for the international 5G research activities. The aim of the ITU-R is to finalize the vision of 5G mobile broadband society which, in turn, is an instrumental base for the ITU's frequency allocation discussions at the WRC events from which the WRC'15 was the most concrete session up today for discussing the 5G frequency strategies. The WRC decides the ways for reorganizing the frequency bands for the current and forthcoming networks, including the ones that will be assigned to the 5G.

Concretely, the Working Party 5D (WP5D) of ITU-R coordinates information sharing about the advances of 5G, including the vision and technical trends, requirements, RF sharing and compatibility, support for applications and deployments, and most importantly, the creation of the IMT-2020 requirement specifications.

ITU-R WP5D uses the same process for 5G as was applied to IMT-Advanced. Specifically for the 5G system evaluation process, the timeline of ITU is the following (Figure 1.3):

- *2016–2017*. Performance requirements, evaluation criteria, and assessment methodology of new radio;

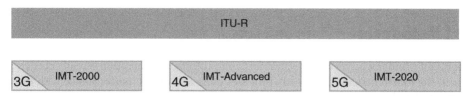

Figure 1.2 The 3G and 4G systems that comply with the ITU requirements for respective generations. The requirements for 5G are also produced by ITU. So far, 3GPP has made a concrete plan to submit the candidate proposal by 2019.

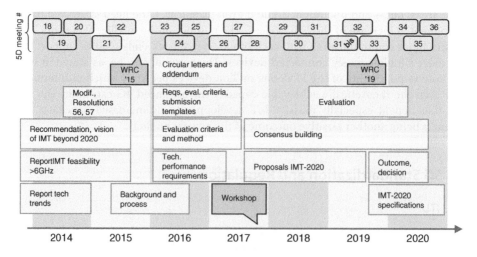

Figure 1.3 ITU time schedule for IMT-2020 as interpreted from [9].

- *2018.* Time frame for proposal;
- *2018–2020.* Definition of the new radio interfaces;
- *2020.* Process completed.

1.9.2 3GPP

While ITU-R is preparing for the evaluation of the 5G candidate technologies that would be compatible with the 5G framework as seen by ITU, one of the active standardization bodies driving the practical 5G solutions is the 3GPP, which is committed to submitting a candidate technology to the IMT-2020 process. The 3GPP is aiming to send the initial technical proposal to the ITU-R WP5D meeting #32 in June 2019 and plans to provide the detailed specification by meeting #36 in October 2020. To align the technical specification work accordingly, the 3GPP has decided to submit the final 5G candidate proposal based on the further evolved LTE-Advanced specifications, as will be their status by December 2019. In addition to the 3GPP, there may also be other candidate technologies seen, such as an enhanced variant of IEEE 802.11.

As for the 3GPP specifications, 5G will affect several technology areas of radio and core networks. The expected aim is to increase the theoretical 4G data rates perhaps 10–50 times higher whilst the response time of the data would be reduced drastically, close to zero. The 3GPP RAN TSG (Radio Access Network Technical Specification Group) is the responsible entity committed to identify more specifically these requirements, scope, and 3GPP requirements for the new radio interface. The RAN TSG works in parallel fashion for enhancing the ongoing LTE evolution that belongs to the LTE-Advanced phase of the 3GPP, aiming to comply with the future IMT-2020 requirements of the ITU. At the same time, the evolved core network technologies need to be revised by the system architecture teams so that they can support the increased data rates accordingly.

3GPP is committed to submitting a candidate technology to the IMT 2020 process based on the following time schedule, as described in the reference SP-150149 (5G timeline in 3GPP) (Figure 1.4):

- *June 2018.* Release 15, Stage 3 freeze;
- *June 2019.* Initial technology submission by ITU-R WP5D meeting #32;
- *October 2020.* Detailed specification submission by ITU-R WP5D meeting #36.

As for the development of the security aspects, 3GPP SA3 has produced the 5G security specification TS 33.501, V15.2.0. Some of the most important topics relevant to UICC follow:

- Tamper-resistant hardware is mandatory for key storage, key derivation, and running the authentication algorithm. Please note that it is not explicitly stated that this applies for both 3GPP and non-3GPP networks and for both primary and secondary authentication.
- Both Extensible Authentication Protocol (EAP) Authentication and Key Agreement (AKA') and 5G AKA are mandatory to be supported for accessing 5G network using a primary authentication.
- 4G and 5G AKA are similar with enhancement on the Authentication Confirmation message.

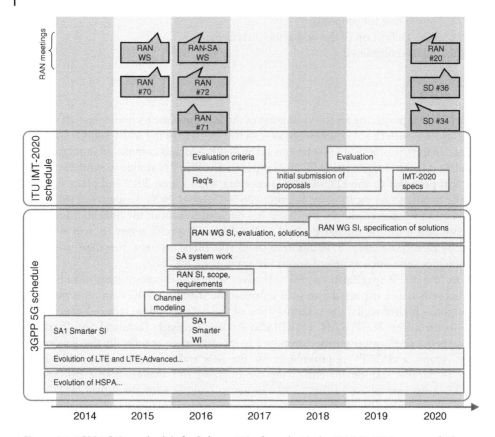

Figure 1.4 3GPP 5G time schedule for Release 15 is aligned with the ITU IMT-2020 progress [10].

- EAP AKA' will also be used to access non-3GPP networks.
- 256-bit algorithms are required in 5G.
- New 5G user identifiers are SUPI (Subscription Permanent Identifier), SUCI (Subscription Concealed Identifier) and 5G-GUTI (5G Globally Unique Temporary Identity).

As for the radio interface, the most remarkable news of 2017 was the decision to include the 3GPP's 5G Next Radio (NR) work item for the non-standalone mode, i.e. for the scenarios with 5G radio base station relying on 4G Evolved Packet Core (EPC). At the same time, there was agreement on:

- Stage 3 for non-standalone 5G-NR evolved Multimedia Broadband (eMBB) includes low-latency support.
- 4G LTE EPC network will be reused.
- The control plane on EPC-eNB-user equipment (UE) will be reused.
- An additional next-generation user plane is adapted on NR gNB–UE.

The 3GPP Release 15 was formally frozen on June 2018, meaning that no new work items were accepted into that release. Release 15 thus contains the first phase of 5G and provides the eMBB services for the early markets, either via non-standalone (NSA) or standalone (SA) modes.

The ASN.1 notation for the NSA was ready on March 2018, while the ASN.1 for the SA variant was ready September 2018. These notation documents are in practice the implementation guides for the equipment manufacturers, and based on these, the first standard-based lightweight 5G networks were deployed soon after.

The second phase of 5G as defined by 3GPP Release 16 with full functionality can be expected to be reality a few months after the freezing of the Release 16 ASN.1 notation set, meaning that the first IMT-2020-compatible 3GPP-based 5G networks will be deployed during 2020. These networks can support also the rest of the 5G pillars in addition to the eMBB, i.e. massive Internet of Things (mIoT) and critical communications referred to as ultrareliable low latency communications (URLLC).

1.10   Global Standardization in 5G Era

The following sections summarize some of the key standardization bodies and industry forums that influence 5G either directly or indirectly, as well as the ones dealing with IoT standardization paving the way for the IoT in the 5G era.

1.10.1   GlobalPlatform

GlobalPlatform is a standardization body with interest areas covering, e.g. UICC and embedded UICC (eUICC), Secure Element (SE), Secure Device, trusted service manager (TSM), certification authority (CA), and TEE. The key standards of the GlobalPlatform related to the IoT include the embedded UICC protection profile, and the body has established an IoT task force. The respective solutions are also valid in 5G, including all the form factors of UICCs (such as embedded and integrated) and their remote management. The organization is outlined in [11] and the respective task forces are listed in [12].

1.10.2   ITU

The International Telecommunications Union (ITU) is a standardization body with a variety of global telecommunications-related requirements and standards. ITU has a leading role in setting the expectations for the 5G era, and the IMT-2020 requirements are the reference for these performance expectations [9, 13, 14].

As for the development of IoT, the ITU has an IoT Global Standards Initiative (IoT-GSI) established in 2015, as well as a study group 20 on IoT, applications, smart cities, and communities. The aim of the ITU is to ensure a unified approach in ITU-T for development of standards enabling the IoT on a global scale. The key IoT-related standard is the Rec. ITU-T Y.2060 (06/2012). It can be expected that the massive IoT will be a major component of the 5G pillars, along with the eMBB and Critical Communications [15]. The ITU-T SG-20 deals with IoT and its applications, including smart cities and communities (SC&C). The resulting standard is designed for IoT and smart cities. There also is an international standard for the development of IoT including M2M communications and sensor networks [16].

The IoT-GSI concluded its activities in July 2015 following Telecommunication Standardization Advisory Group's (TSAG) decision to establish the new Study Group 20 on *IoT and its applications including smart cities and communities.* All activities ongoing in the IoT-GSI were transferred to the SG20. The IoT-GSI aimed to promote a unified approach in ITU-T for the development of technical standards enabling the IoT on a global scale. ITU-T recommendations developed under the IoT-GSI by the various ITU-T questions, in collaboration with other standards developing organizations (SDOs), will enable worldwide service providers to offer the wide range of services expected by this technology. The IoT-GSI also acts as an umbrella for IoT standards development worldwide [17].

1.10.3 IETF

The Internet Engineering Task Force (IETF) develops the Internet architecture. The IETF has dedicated IETF Security Area (https://tools.ietf.org/area/sec/trac/wiki). Some of the key standards include the Constrained Application Protocol (CoAP), and adaptation to the current communication security for use with CoAP. There also is the standard for IPv6 Low-power Wireless Personal Area Network (6LoWPAN) [18]. The IoT directory of IETF is found in [19].

1.10.4 3GPP/3GPP2

3GPP and its US counterparty 3GPP2 focus on cellular connectivity specifications that have been actively widened to cover also low-power wide area network (LPWAN) area. These include most concretely LTE-M and category 0 for low-bit rate M2M, and the further enhanced terminal categories that are optimized for IoT such as Cat-M1 and NB-IoT. The standardization body also develops security aspects, 2G/3G/4G/5G security principles and architectures, algorithms, lawful interception, key derivation, backhaul security, and SIM/UICC that can give added value for the cellular IoT compared to the competing proprietary variants [20].

1.10.5 ETSI

The European Telecommunications Standards Institute (ETSI) executes security standardization of, e.g. UICC and its evolution under the term SSP (smart secure platform). The latter is a continuum that opens room for new forms of UICCs such as embedded and integrated UICCs. The ETSI Technical Committee (TC) M2M is a relevant group for IoT development at ETSI. The organization's overall information can be found at Ref. [21] and IoT-related program in [22]. Furthermore, the ETSI portal of work documents for members is found at [23].

1.10.6 IEEE

The Institute of Electrical and Electronics Engineers (IEEE) 802 series includes aspects for IoT connectivity. There also exist many other related standards useful for IoT environment, like the IEEE Std 1363 series for public key cryptography. The IEEE 802.11 has many variants from which e.g. 802.11p is designed specifically for vehicle-to-vehicle

(V2V) communications. That can be considered as a competing technology for the 3GPP-based modes that will be optimized for vehicle communications, especially in the 5G era [24].

The IEEE Project P.2413 revises IEEE standards for better use within the IoT. The goal of the project is to build reference architecture covering the definition of basic architecture building blocks and integration into multitiered systems. The architectural framework for IoT provides a reference model that defines relationships among various IoT verticals, including transportation and health care, as well as common architecture elements. It also provides means for data abstraction, protection, security, privacy, and safety. The reference architecture of the project covers the basic architectural building blocks and their ability to be integrated into multitiered systems [25].

1.10.7 SIMalliance

The task of SIMalliance is to simplify SE implementation, and it drives deployment and management of secure mobile services. It also promotes SE for secure mobile applications and services and promotes subscription management standardization, which is beneficial to provide a standardized means for the remote management of embedded universal integrated circuit card (eUICC) and integrated universal integrated circuit card (iUICC), which can be expected to be elemental components of 5G ecosystem [26].

1.10.8 Smart Card Alliance

Smart Card Alliance (SCA) has been a centralized industry interface for smart card technology, and it has followed the impact and value of smart cards in the United States and Latin America. As the 5G IoT can be based largely on the basic concept of the SIM card, and thus smart card technology, this task continues being relevant in the 5G era as well. From its inception as the SCA, its current form of Secure Technology Alliance (STA) facilitates the adoption of secure solutions in the United States. The Alliance's focus is set on securing a connected digital world by driving adoption of new secure solutions [27].

1.10.9 GSMA

The GSM Association (GSMA) represents interests of MNOs worldwide. It is involved in the Network 2020 paving the way for 5G. GSMA is involved in the standardization of subscription management and embedded Subscriber Identity Module (eSIM), and their development for M2M and consumer environment. It should be noted that the previous term of GSMA for indicating remote Subscriber Identity Module provisioning (RSP) is now generalized via the term eSIM, which has been approved by the GSMA as a global product label that can be used to indicate that a device is "RSP enabled" [28].

1.10.10 NIST

The US National Institute of Standards Technology (NIST) develops cybersecurity frameworks to address critical infrastructure including IoT/M2M space. The focus is on security and privacy for the evolution of IoT and M2M. It produces Federal

Information Processing Standards Publications (FIPS PUBS). FIPS are developed by the Computer Security Division within the NIST for protecting federal assets such as computer and telecom systems. FIPS 140 (1–3) contains security requirements. The topics on International Technical Working Group on IoT-Enabled Smart City Framework of NIST are found in [29] and FIPS PUBS are in [30].

1.10.11   NHTSA

The National Highway Transportation and Safety Administration (NHTSA) improves safety and mobility on US roadways. It also investigates connected vehicle technology and communications of safety and mobility information to one another. It has International Technical Working Group on IoT-Enabled Smart City Framework [31].

1.10.12   ISO/IEC

The International Organization for Standardization (ISO)/International Electrotechnical Commission (IEC) is an elemental body for smart card technology standardization. ISO/IEC 7816 and 14400 are SIM/UICC standards for contact-oriented and contactless integrated circuit cards (ICCs). There are various solutions in the markets based on these standards, including transport cards, and IoT devices can be expected to be based largely on UICC. ISO/IEC 27000 is Information Security Management framework, which is valid also for IoT security.

Related to IoT security, the ISO/IEC Common Criteria (CC) is an international security evaluation framework that provides reliable IT product evaluation for the security capabilities based on an international standard (ISO/IEC 15408) for computer security certification, which refers to standards denoting EAL (evaluation assurance level) of 1–7. ISO/IEC 19794 produces biometrics standards [32].

1.10.13   ISO/IEC JTC1

Joint Technical Committee (JTC) 1 is the standards development environment to develop worldwide information and communication technology (ICT) standards for business and consumer applications. Additionally, JTC 1 provides the standards approval environment for integrating diverse and complex ICT technologies. ISO/IEC JTC 1/SC 27 deals with IT security techniques [33].

1.10.14   OMA

Open Mobile Alliance (OMA) has developed device management (DM). OMA LightweightM2M aims to optimize the secure communications between all, especially economic, devices. OMA DM is a subgroup under the OMA alliance. OMA DM is an initiative for automotive environment, and it includes over the air (OTA) updates for future investigations. The role of OMA is detailed in [34].

1.10.15 CEPT/ECC

Conférence Européenne des Postes et des Télécommunications (CEPT) and Electronic Communications Committee (ECC) are coordinated by European Communications Office (ECO). They produce requirements for approval for certification bodies and testing labs. They work in the ECC on smart grids, smart metering, and others under the ultra-high frequency (UHF) roadmap. Related to IoT environment, there is an ECC Report 153 on Numbering and Addressing in M2M Communications.

M2M can be used in several licensed and unlicensed frequency bands. The aim of ECC is to understand better the spectrum as well as numbering and addressing harmonization needs of existing and future M2M applications since related initiatives are present in various for a within the ECC, aligning with industry. The Working Group for Frequency Management (WGFM) of the ECC had prepared information to the ECC on the regulatory framework for M2M communications on the basis of frequency bands already available for various M2M usages; see ECC(15)039 Annex 13 [35].

1.10.16 NERC

Indirectly related to IoT, North American Electric Reliability Corporation (NERC) is committed to protecting the bulk power system against cybersecurity compromises that could lead to faulty operation or instability. CIP refers to Critical Infrastructure Protection cybersecurity standards, the CIP V5 Transition Program being the most recent one in the United States [36].

1.10.17 OWASP

Open Web Application Security Project (OWASP) is a worldwide not-for-profit charitable organization focused on improving the security of software. OWASP IoT Project provides information on IoT attack surface areas and IoT testing guides and maintains a top-10 IoT vulnerabilities list [37].

1.10.18 OneM2M

OneM2M's architecture and standards for M2M communications are designed to be applied in many different industries and take account of input and requirements from any sector. It works on eHealth and Telemedicine, Industrial, and Home Automation [38].

1.10.19 Global Standards Collaboration

Global Standards Collaboration (GSC) is an unincorporated voluntary organization dedicated to enhancing global cooperation and collaboration regarding communications standards and the related standards development environment. GSC is not a standards development organization and therefore will not develop standards. The members of GSC include ARIB (Association of Radio Industries and Businesses in Japan), ATIS (Alliance for Telecommunications Industry Solutions in the United States), CCSA (China Communications Standards Association), ETSI, IEC, IEEE-SA (IEEE

Standards Association), ISO, ITU, TIA (Telecommunications Industry Association in the United States), TSDSI (Telecommunications Standards Development Society in India), TTA (Telecommunications Technology Association in Korea), and TTC (Telecommunication Technology Committee in Japan) [39].

1.10.20 CSA

Cloud Security Alliance (CSA) is a nonprofit organization to promote the use of practices for providing security assurance within cloud computing and provide education on the uses of cloud computing to help secure all other forms of computing. CSA operates the cloud security provider certification program, the CSA Security, Trust & Assurance Registry (STAR), a three-tiered provider assurance program of self-assessment, third-party audit, and continuous monitoring [40]. Cloud computing and data centers form an integral part of the 5G infrastructure.

1.10.21 NGMN

Next Generation Mobile Networks (NGMN) is relevant for overall advanced network technologies as well as for the IoT [41]. As an example, NGMN has launched a projects "Spectrum and deployment efficiencies," "URLLC requirements for vertical industries," "RAN convergence," and "Extreme long-range communications for deep rural coverage." These activities are aimed to optimize and guide the telecoms industry toward the successful deployment of 5G beyond 2018.

1.10.22 Car-to-Car Communication Consortium

Car-to-Car Communication Consortium (C2C-CC) is industry forum for the V2V technology development. It is a nonprofit, industry driven organization initiated by European vehicle manufacturers and supported by equipment suppliers, research organizations, and other partners. The C2C-CC is dedicated to the objective of further increasing road traffic safety and efficiency by means of cooperative intelligent transport systems (C-ITS) with V2V communication supported by vehicle-to-infrastructure communication (V2I). It supports the creation of European standards for communicating vehicles spanning all brands. As a key contributor, the C2C-CC works in close cooperation with European and international standardization organizations [42].

1.10.23 5GAA

The mission of the 5G Automotive Association (5GAA) is to develop, test, and promote communications solutions, initiate their standardization, and accelerate their commercial availability and global market penetration to address society's connected mobility and road safety needs with applications such as autonomous driving, ubiquitous access to services, and integration into smart city and intelligent transportation. 5GAA offers several levels of membership for corporations, industry organizations, and academic institutions [43].

1.10.24   Trusted Computing Group

Through open standards and specifications, Trusted Computing Group (TCG) enables secure computing. Some benefits of TCG technologies include protection of business-critical data and systems, secure authentication and protection of user identities, and the establishment of machine identity and network integrity. The work groups are cloud; embedded systems; infrastructure; IoT; mobile; PC client; regional forums; server; software stack; storage; trusted network communications; trusted platform module (TPM); and virtualized platform [44].

1.10.25   InterDigital

InterDigital, Inc. designs and develops advanced technologies that enable and enhance mobile communications and capabilities for concept of the Living Network basing on intelligent networks, which self-optimize to deliver services tailored to the content, context and connectivity of the user, device, or need. Ecosystem partners providing devices, platforms, and data services, and affiliations include OneM2M, Industrial Internet Consortium, and Convida [45].

1.11   Introduction to the Book

This book is designed for the technical personnel of operators, equipment manufacturers, and telecom students. Previous knowledge about mobile communications would help in capturing the most detailed messages of the book, but the modular structure of the chapters – including the introductory part for the technology – is aimed to ensure that the book is useful also for the readers who are not yet familiar with the subject.

The readers in mobile equipment engineering, security, network planning, and optimization, as well as application development teams, benefit from the contents, as it details highly novelty aspects of the field, helping in updating the essential information in a compact way. The book is meant primarily for specialists of the field with a need to quickly capture the new key aspects of 5G, important differences compared to previous generations, and the possibilities and challenges in the 5G network deployment.

This contents thus demystifies the idea of fifth-generation mobile communications basing on the latest 5G standards and summarizes available information into a compact book form. The book focuses especially on the security aspects as well as on the network planning and deployment of the forthcoming 5G. It summarizes the 5G functionality with a special emphasis on the new security requirements. It discusses the security techniques and gives common-sense guidelines for planning, optimizing, and deploying the networks.

Chapters 1–3 of this book form the introductory module that will be useful for both technical and nontechnical readers with or without preliminary knowledge about existing mobile communications systems. Chapters 4–7 form the technical description and are directed to the advanced readers with some knowledge on mobile communications, while Chapters 8–10 represent the planning module and are meant for seasonal subject matter experts.

Figure 1.5 presents the main-level contents of this book to ease navigation between the modules. The modules and chapters are independent from each other, so they can be

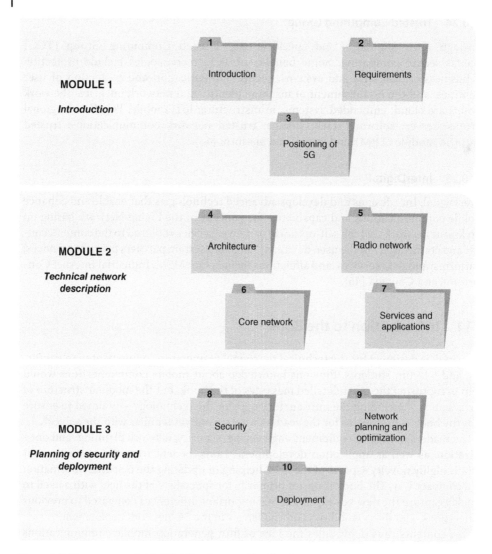

MODULE 1

Introduction

1 Introduction

2 Requirements

3 Positioning of 5G

MODULE 2

Technical network description

4 Architecture

5 Radio network

6 Core network

7 Services and applications

MODULE 3

Planning of security and deployment

8 Security

9 Network planning and optimization

10 Deployment

Figure 1.5 The contents of this LTE-A Deployment Handbook.

read through in any preferred order. Nevertheless, if you are starting from scratch (i.e. your knowledge of the subject area is less advanced), it is recommended that chapters be read in chronological order.

References

1 Penttinen, J. (2016). *The LTE-Advanced Deployment Handbook*. London: Wiley.
2 Penttinen, J. (2016). *The Wireless Communications Security*. London: Wiley.
3 Penttinen, J. (2015). *The Telecommunications Handbook*. London: Wiley.

4 Verizon, "Verizon 5G Technical Forum," Verizon, http://5gtf.net. [Accessed 3 July 2018].

5 Qualcomm, "Making 5G NR mmWave a commercial reality," Qualcomm, 2018.

6 European Commission, "Towards 5G," 16 August 2017. http://ec.europa.eu/digital-agenda/en/towards-5g. [Accessed 26 September 2017].

7 European Union, "EU and Brazil to work together on 5G mobile technology," 23 February 2016. http://europa.eu/rapid/press-release_IP-16-382_en.htm. [Accessed 26 September 2017].

8 NGMN, "NGMN 5G White Paper," NGMN, 2015.

9 ITU, "ITU towards "IMT for 2020 and beyond"," ITU, http://www.itu.int/en/ITU-R/study-groups/rsg5/rwp5d/imt-2020/Pages/default.aspx. Accessed 3 July 2018.

10 3GPP, "Tentative 3GPP timeline for 5G," 3GPP, 17 March 2015. http://www.3gpp.org/news-events/3gpp-news/1674-timeline_5g. [Accessed 3 July 2018].

11 GlobalPlatform, "GlobalPlatform," GlobalPlatform, https://www.globalplatform.org. Accessed 3 July 2018.

12 GlobalPlatform, "GlobalPlatform," GlobalPlatform, https://www.globalplatform.org/aboutustaskforcesIPconnect.asp. [Accessed 3 July 2018].

13 E. Mohyeldin, "Minimum Technical Performance Requirements for IMT-2020 radio interface(s)," ITU, 2016.

14 M. Carugi, "Key features and requirements of 5G/IMT-2020 networks," ITU, Algeria, 2018.

15 ITU, "ITU," ITU, www.itu.int. Accessed 3 July 2018.

16 ITU, "ITU IoT," ITU, http://www.itu.int/en/ITU-T/studygroups/2013-2016/20/Pages/default.aspx. [Accessed 3 July 2018].

17 ITU, "Internet of Things Global Standards Initiative," ITU, http://www.itu.int/en/ITU-T/gsi/iot/Pages/default.aspx. [Accessed 4 July 2018].

18 IETF, "IETF," IETF, https://www.ietf.org. [Accessed 4 July 2018].

19 IETF, "Internet-of-Things Directorate," IETF, https://trac.ietf.org/trac/int/wiki/IOTDirWiki. Accessed 4 July 2018.

20 3GPP, "3GPP," 3GPP, www.3gpp.org. Accessed 4 July 2018.

21 ETSI, "ETSI," ETSI, www.etsi.org. Accessed 3 July 2018.

22 ETSI, "Connecting Things," ETSI, http://www.etsi.org/technologies-clusters/clusters/connecting-things. [Accessed 3 July 2018].

23 ETSI, "ETSI work documents," ETSI, https://portal.etsi.org/tb.aspx?tbid=726&SubTB=726. Accessed 3 July 2018.

24 IEEE, "Internet of Things," IEEE, http://iot.ieee.org. Accessed 03 07 2018.

25 IEEE, "IEEE Project 2413 – Standard for an Architectural Framework for the Internet of Things (IoT)," IEEE, https://standards.ieee.org/develop/project/2413.html. [Accessed 03 07 2018].

26 SIMalliance, "Identity, security, mobility," SIMalliance, http://simalliance.org. [Accessed 3 July 2018].

27 Secure Technology Alliance, "Digital security industry's premier association," Secure Technology Alliance, http://www.smartcardalliance.org. [Accessed 3 July 2018].

28 GSMA, "Representing the worldwide mobile communications industry," GSMA, http://www.gsma.com. [Accessed 3 July 2018].

29 NIST, "International Technical Working Group on IoT-Enabled Smart City Framework," NIST, https://pages.nist.gov/smartcitiesarchitecture. [Accessed 3 July 2018].

30 NIST, "Federal Information Processing Standards Publications (FIPS PUBS)," NIST, https://www.nist.gov/itl/popular-links/federal-information-processing-standards-fips. [Accessed 3 July 2018].

31 NHTSA, "Main page," NHTSA, http://www.nhtsa.gov. [Accessed 3 July 2018].

32 ISO, "International Organization for Standardization," ISO, www.iso.org. [Accessed 3 July 2018].

33 ISO, "ISO/IEC JTC 1 – Information Technology," ISO, http://www.iso.org/iso/jtc1_home.html. Accessed 3 July 2018.

34 OMA, "OMA SpecWorks," OMA, http://openmobilealliance.org. [Accessed 3 July 2018].

35 CEPT, "Electronic Communications Committee," CEPT, http://www.cept.org/ecc. [Accessed 3 July 2018].

36 NERC, "CIP V5 Transition Program," NERC, http://www.nerc.com/pa/CI/Pages/Transition-Program.aspx. Accessed 3 July 2018.

37 OWASP, "The OWASP Foundation," OWASP, https://www.owasp.org/index.php/Main_Page. [Accessed 3 July 2018].

38 OneM2M, "Standards for M2M and the Internet of Things," OneM2M, http://www.onem2m.org. [Accessed 4 July 2018].

39 Global Standards Collaboration, "Global Standards Collaboration," ITU, 04 07 2018. [Online]. Available: http://www.itu.int/en/ITU-T/gsc/Pages/default.aspx. Accessed 4 July 2018.

40 Cloud Security Alliance, "Cloud Security Alliance," Cloud Security Alliance, https://cloudsecurityalliance.org. Accessed 4 July 2018.

41 NGMN, "NGMN," NGMN, www.ngmn.org. Accessed 4 July 2018.

42 CAR 2 CAR Communication Consortium, "CAR 2 CAR Communication Consortium," CAR 2 CAR Communication Consortium, https://www.car-2-car.org. Accessed 4 July 2018.

43 5GAA, "5GAA," 5GAA, http://5gaa.org. Accessed 4 July 2018.

44 Trusted Computing Group, "Trusted Computing Group," Trusted Computing Group, https://trustedcomputinggroup.org. Accessed 4 July 2018.

45 InterDigital, "Creating the living network," InterDigital, http://www.interdigital.com/page/about. Accessed 4 July 2018.

2

Requirements

2.1 Overview

This chapter summarizes the technical requirements for fifth-generation (5G) systems. It should be noted that this book assumes ITU (International Telecommunications Union) to be the highest authority dictating the minimum 5G requirements. This book thus walks the reader through the explanation of the latest ITU statements for their 5G candidate selection. Furthermore, this chapter summarizes the statements the other relevant standardization body has presented, the Third Generation Partnership Project (3GPP) being one of the most active stakeholders in the standardization of the 5G.

The time schedule for the key aspects of the 5G development and deployment, by the writing of this publication, includes the ITU-R's (the radio section of ITU) high-level requirements presented in the document [1], and its finalized version [2]. The concrete requirement set for the candidate evaluation have been presented by the end of 2017 as stated in [3].

The 3GPP, on the other hand, is one of the most relevant standardization bodies to present a candidate 5G technology for ITU's 5G selection process. The 3GPP has sent the first draft technical specifications as defined in the Release 15 for the preliminary ITU-R review, and the final submission will be based on the frozen technical standard set by Release 16.

Other candidate technologies might also be presented to the ITU evaluation, according to the ITU time schedule, although as an example, the Institute of Electrical and Electronics Engineers (IEEE) has not planned to send a complete 5G system proposal to the ITU evaluation; instead, there will be many IEEE sub-solutions that are relevant in building up 5G networks.

This chapter thus introduces new requirements for 5G as far they can be interpreted from various relevant sources. These key sources of information are presented as available now, and based on this information, the requirements are interpreted for the already known and foreseen statements, e.g. via ITU-R, 3GPP, and other entities involved with the standardization of 5G.

This chapter also presents the aspects of the interworking of 5G services with legacy systems, presenting general considerations of cooperative functioning of 5G and other relevant systems. Also, the performance aspects of the 5G in fixed and wireless environment are discussed, along with the possibilities and constraints, including the performance in practice now and via expected standardized features. Finally,

5G Explained: Security and Deployment of Advanced Mobile Communications, First Edition.
Jyrki T. J. Penttinen.
© 2019 John Wiley & Sons Ltd. Published 2019 by John Wiley & Sons Ltd.

there is discussion on the impacts of requirements, including impact analysis for the technologies and businesses.

2.2 Background

With such a variety of 5G announcements for the trial phase as well as the expected 5G deployments, and each press release indicating only limited functionalities and early, preliminary (i.e. not yet International Mobile Telecommunications [IMT]-2020 – compliant) approaches, one might ask who will dictate how 5G should look? There is obviously great interest in the industry to maintain the competitive edge compared to other stakeholders whilst the global standardization and jointly approved and agreed principles of 5G are still under work.

As is customary, ITU has taken the role of defining the mobile communications generations. This was the situation for the third generation (3G) and fourth generation (4G), after the success of the first, analogue generation, and second, digital generations. ITU defines 3G as a set of radio access and core technologies forming systems capable of complying with the IMT-2000 (3G) and International Mobile Telecommunications Advanced (IMT-Advanced) (4G) requirements. Based on the number of end users, Universal Mobile Telecommunications System (UMTS) and its evolution up to advanced HSPA (high-speed packet access) is the most popular 3G system while Long-Term Evolution (LTE)-Advanced, as of 3GPP Release 10, is the most utilized 4G technology.

As for the industry, the Next Generation Mobile Network (NGMN) Alliance represents the interests of mobile operators, device vendors, manufacturers, and research institutes. The NGMN is an open forum for participants to facilitate the evaluation of candidate technologies suitable for the evolved versions of wireless networks. One of the main aims of the forum is to pave the way for the commercial launch of new mobile broadband networks. Some practical methods to do this are the production of commonly agreed technology roadmap as well as user trials [4].

So, ITU is still acting as a highest authority to define the global and interoperable 5G requirements for the mobile communications systems in such a way that the requirements are agreed by all the stakeholders. This is to avoid different interpretations by the industry when deploying and marketing the networks.

ITU setting the scenery, the practical standardization work results in the ITU-5G-compliant technical specifications created by mobile communications industry. One of the most active standardization bodies for the mobile communications technologies is 3GPP, which has created standards for Global System for Mobile communications (GSM), UMTS/HSPA, and LTE/LTE-Advanced. At present, 3GPP is actively creating advanced standards that are aimed to comply with ITU's 5G requirements.

In addition to the 3GPP, many other standardization bodies and industry forums contribute to the 5G technologies. Maybe not complete end-to-end 5G systems are under construction in such a large scale, as is done by 3GPP, but, e.g. many recommendations by IEEE are used as a base for 5G (as well as any other existing mobile communication technology). The new IEEE recommendations are formalizing the 5G, as one part of the complete picture.

2.3 5G Requirements Based on ITU

2.3.1 Process

The ITU Recommendation ITU-R M.2083 describes the IMT-2020 overall aspects. It will provide enhanced capabilities, which are much more advanced compared to the ones found in the ITU Recommendation ITU-R M.1645. The IMT-2020 has a variety of aspects from different points of view, which greatly extends the requirements compared to previous mobile communication generations.

The ecosystem and respective performance are related to users, manufacturers, application developers, network operators, as well as service and content providers. This means that there are many deployment scenarios supported by the IMT-2020 with a multitude of environments, service capabilities, and technological solutions [1].

The ITU represents the highest authority in the field of defining the mobile system generations.

The overall vision of 5G, according to the ITU, is presented in [5]. In short, the ITU foresees the 5G to function as an enabler for a seamlessly connected society in the 2020 time frame and beyond. The high-level idea of the 5G is to bring together people via a set of "things," data, applications, transport systems, and cities in a smart networked communications environment. The ITU and the respective, interested partners believe that the relationship between IMT and 5G are elements that make it possible to deploy the vision in practice by relying on mobile broadband communications.

The idea of 5G was presented as early as 2012 when ITU-R initiated a program to develop IMT for the year 2020 and beyond. As a result, there were research activities established in global environment.

The ITU Working Party 5D (WP5D) has been working on the ITU's expected time frame paving the way for the IMT-2020, including the investigation of the key elements of 5G in cooperation with the mobile broadband industry and other stakeholders interested in the 5G.

One of the elemental parts of this development has been the 5G vision of the ITU-R for the mobile broadband connected society, which was agreed in 2015. This vision is considered as instrumental and serves as a solid foundation for the World Radiocommunication Conference 2019. That will be a major event for the decisions of the 5G frequency bands, including the additional spectrum that will be required in different regions for the massively increasing mobile communications traffic.

ITU has played an essential role in the development of mobile radio interface standards. The previous requirements of the ITU ensuring the internationally recognized systems for 3G (IMT-2000) and 4G (IMT-Advanced) is now extending to cover 5G via the IMT-2020 requirements. Figure 2.1 depicts the time schedule of the ITU for the development of the 5G, and for paving the way for the very first deployments as of 2020.

From various teams considering 5G evolution, one of the most significant is the ITU-R WP5D. It has a role of investigating study areas and deliverables toward IMT for 2020 and beyond via multitude of activities, such as workshops and seminars for the information sharing within the industry and standardization entities. Some of the more specific work item categories include the following:

- *Vision and technology trends.* These aspects include also market, traffic, and spectrum requirements for the forthcoming 5G era.

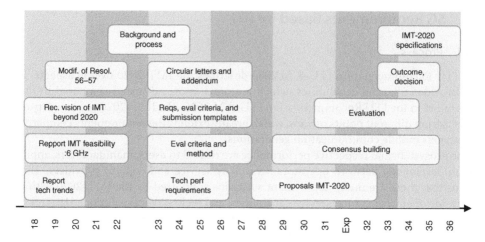

Figure 2.1 The main phases of the IMT-2020 development by the ITU. Within this process, World Radio Conferences 2015 and 2019 play an essential role for extending the 5G frequency band utilization.

- *Frequency band.* Investigations focus on frequency band channeling arrangements and spectrum sharing and compatibility.
- *IMT specifications.* This item also contains related technical works.
- *Support for IMT applications and deployments.* This item addresses the current and future use of International Mobile Telecommunications, such as the LTE to support broadband public protection and disaster relief (PPDR) communications.

The ITU IMT-2020 requirements form the foundation of the internationally recognized 5G systems. The respective ITU Radiocommunications Bureau Circular Letters work for announcing the invitation for standardization bodies to submit their formal 5G technical proposals for the ITU Working Party 5D evaluation of their compliance with the IMT-2020 requirements. This process follows the principles applied in the previous IMT-Advanced for selecting the 4G systems. Prior to the candidate evaluation, the WP5D finalized the performance requirements 2017 and formed the evaluation criteria and methodology for the assessment of the IMT-2020 radio interface.

After the candidate submission, the WP5D evaluates the proposals during the time of 2018–2020. The work is based on independent, external evaluation groups, and the process is estimated to be completed during 2020, along with draft ITU-R Recommendation that contains detailed specifications for the new radio (NR) interfaces.

2.3.2 Documents

The key sources of information for the ITU 5G development can be found in ITU-R M.2083 [6], which collects the documents for forming the 5G:

- *M.2083-0 (09/2015). IMT vision.* "Framework and overall objectives of the future development of IMT for 2020 and beyond." This document contains the recommendations for the future development of IMT for 2020 and beyond, and it defines the framework and overall objectives of the future development considering the roles that IMT could play to better serve the needs of the networked societies. It also includes a variety of detailed capabilities related to foreseen usage scenarios, as well as objectives of the future development of IMT-2020 and existing IMT-Advanced. It is based on Recommendation ITU-R M.1645.

- *Report ITU-R M.2320*. Future technology trends of terrestrial IMT systems addresses the terrestrial IMT technology aspects and enablers during 2015–2020. It also includes aspects of terrestrial IMT systems related to WRC-15 studies.
- *Report ITU-R M.2376*. Technical feasibility of IMT in bands above 6 GHz summarizes the information obtained from the investigations related to the technical feasibility of IMT in the bands above 6 GHz as described in ITU-R Rec. 23.76 [7].
- *Recommendation ITU-R M.1645*. Framework and overall objectives of the future development of IMT-2000 and systems beyond IMT-2000.
- *Recommendation ITU-R M.2012*. Detailed specifications of the terrestrial radio interfaces of IMT-Advanced.
- *Report ITU-R M.2370*. IMT Traffic estimates for the years 2020–2030.
- *Report ITU-R M.2134*. Requirements related to technical performance for IMT-Advanced radio interface(s).

As summarized in the press release [8], ITU agreed the first 5G requirement set on 23 February 2017. The ITU's IMT-2020 standardization process follows the schedule plan presented in Figure 2.2.

ITU has published the minimum 5G requirements in the document IMT-2020 Technical Performance Requirements, available at [1]. The most important minimum set of technical performance requirements defined by ITU provide means for consistent definition, specification and evaluation of the candidate IMT-2020 radio interface technologies (RITs), or a set of radio interface technologies (SRITs).

There is a parallel, ongoing development for the ITU-R recommendations and reports related to 5G, including the detailed specifications of IMT-2020. As a highest-level global authority of such requirements for a new mobile communications generation,

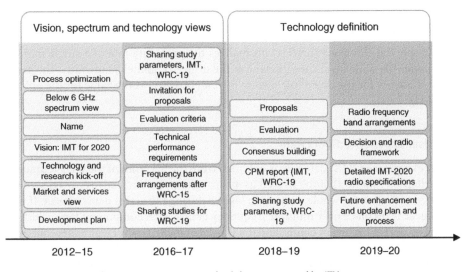

Figure 2.2 The high-level IMT-2020 process schedule as presented by ITU.

ITU aims, with the production of these requirements, to ensure that IMT-2020 technologies can comply with the objectives of the IMT-2020. Furthermore, the requirements set the goal for the technical performance that the proposed set of RITs must achieve for being called ITU-R's IMT-2020 compliant 5G technologies.

ITU will evaluate the IMT-2020-compliant 5G candidate technologies based on the following documents and for the development of IMT-2020:

- Report ITU-R IMT 2020 M-Recommendation for Evaluation
- Report ITU-R IMT-2020 M-Recommendation for Submission

The Recommendation ITU-R M.2083 contains eight *key capabilities* for IMT-2020, and functions as a basis for the technical performance requirements of 5G. It should be noted that the key capabilities have varying relevance and applicability as a function of different use cases within the IMT-2020.

In summary, the ITU's minimum radio interface requirements include the 5G performance requirements is presented in the next section. More specific test evaluation is described in the IMT-2020 Evaluation report of ITU-R [9].

2.3.3 Peak Data Rate

The peak data rate (b/s) refers to the maximum possible data rate assuming ideal, error-free radio conditions that are assigned to a single mobile station with all the available radio resources, excluding the resources for physical layer synchronization, reference signals, pilots, guard bands, and guard times. The term W representing bandwidth, E_{sp} referring to a peak spectral efficiency, the user's peak data rate $Rp = WE_{sp}$. The total peak spectral efficiency is obtained by summing the value per each applicable component frequency bandwidth. This requirement is meant to evaluate the evolved Multimedia Broadband (eMBB) use case, for which the minimum downlink peak data rate is 20 Gb/s whereas the value for uplink is 10 Gb/s.

2.3.4 Peak Spectral Efficiency

The peak spectral efficiency (b/s/Hz) normalizes the peak data rate of a single mobile station under the same ideal conditions over the utilized channel bandwidth. The peak spectral efficiency for the downlink is set to 30 b/s/Hz, whereas the value for the uplink is 15 b/s/Hz.

2.3.5 User Experienced Data Rate

The user experienced (UX) data rate is obtained from the 5% point of the CDF (cumulative distribution function) of the overall user throughput, i.e. correctly received service data units (SDUs) of the whole data set in layer 3 during the active data transfer. If the data transfer takes place over multiple frequency bands, each component bandwidth is summed up over the relevant bands, and the UX data rate $R_{user} = WE_{s\text{-}user}$. This equation refers to the channel bandwidth multiplied by the fifth percentile user spectral efficiency. The ITU requirements for the UX data rate in downlink is 100 Mb/s, whereas it is 50 Mb/s for uplink.

2.3.6 Fifth Percentile User Spectral Efficiency

The fifth percentile user spectral efficiency refers to the 5% point of the CDF of the normalized user data throughput. The normalized user throughput (b/s/Hz) is the ratio of correctly received SDUs in layer 3 during selected time divided by the channel bandwidth. This requirement is applicable to eMBB use case, and the requirement values are, for downlink and uplink, respectively, the following:

- *Indoor hotspot.* 0.3 b/s/Hz (DL) and 0.21 b/s/Hz (UL).
- *Dense urban.* 0.225 b/s/Hz (DL) and 0.15 b/s/Hz (UL), applicable to the Macro TRxP (Transmission Reception Point) layer of Dense Urban eMBB test environment.
- *Rural.* 0.12 b/s/Hz (DL) and 0.045 b/s/Hz (UL), excluding the LMLC (low mobility large cell) scenario.

2.3.7 Average Spectral Efficiency

Average spectral efficiency can also be called spectrum efficiency, as has been stated in ITU Recommendation ITU-R M.2083. It refers to the aggregated throughput, taking into account the data streams of all the users. More specifically, the spectrum efficiency is calculated via the correctly received SDU bits on layer 3 during a measurement time window compared to the channel bandwidth of a specific frequency band divided further by the number of TRxPs, resulting in the value that is expressed in b/s/Hz/TRxP. The ITU requirement values are, for downlink and uplink for eMBB use case, respectively, the following:

- *Indoor hotspot.* 9 b/s/Hz/TRxP (DL) and 6.75 b/s/Hz/TRxP (UL).
- *Dense urban for macro TRxP layer.* 7.8 b/s/Hz/TRxP (DL) and 5.4 b/s/Hz/TRxP (UL).
- *Rural (including LMLC).* 3.3 (DL) and 1.6 (UL).

2.3.8 Area Traffic Capacity

The area traffic capacity refers to the total traffic throughput within a certain geographic area, and is expressed in $Mb/s/m^2$. More specifically, the throughput refers to the correctly received bits in layer 3 SDUs during a selected time window. If the bandwidth is aggregated over more than one frequency band, the area traffic capacity is a sum of individual bands. The target value for the area traffic capacity is set to 10 $Mb/s/m^2$.

2.3.9 Latency

The user plane latency refers to the time it takes for the source sending a packet in radio protocol layer 2/3 to reach its destination on the respective layer. The latency is expressed in milliseconds. The requirement for the user plane latency is 4 ms for eMBB use case, whilst it is 1 ms for ultra-reliable low latency communications (URLLCs). The assumption here is an unloaded condition in both downlink and uplink without other users than the observed one whilst the packet size is small (zero payload and only internet protocol [IP] header).

The control plane latency, in turn, refers to the transition time it takes to change from the idle stage to the active stage in URLLC use case. The requirement for the control plane latency is maximum of 20 ms, and preferably 10 ms or less.

2.3.10 Connection Density

Connection density refers to the total number of 5G devices that can still comply with the target QoS (quality of service) level within a geographical area which is set to $1\,km^2$, with a limited frequency bandwidth and the number of the TRxPs, the variables being the message size, time and probability for successful reception of the messages. This requirement applies to the massive machine-type communications (mMTCs) use case, and the minimum requirement for the connection density is set to $1\,000\,000$ devices per km^2.

2.3.11 Energy Efficiency

The high-level definition of the 5G energy efficiency indicates the capability of the radio interface technology (RIT) and set of RITs (SRIT) to minimize the radio access network (RAN) energy consumption for the provided area traffic capacity. Furthermore, the device energy efficiency is specifically the capability of the RIT and SRIT to optimize the consumed device modem power down to a minimum that still suffices for the adequate quality of the connection. For the energy efficiency of the network as well as the device, the support of efficient data transmission is needed for the loaded case, and the energy consumption should be the lowest possible for the cases when data transmission is not present. For the latter, the sleep ratio indicates the efficiency of the power consumption. Energy efficiency is relevant for the eMBB use case, and the RIT and SRIT must have the capability to support a high sleep ratio and long sleep duration.

2.3.12 Reliability

The reliability of 5G in general refers to the ability of the system to deliver the desired amount of packet data on layers 2 and 3 within expected time window with high success probability, which is dictated by a channel quality. This requirement is applicable to the URLLC use cases.

More specifically, the reliability requirement in 5G has been set to comply with the successful reception of a 32-bit packet data unit (PDU) on layer 2 within $1\,ms$ period with a 1×10^{-5} success probability. This requirement is applicable to the edge of the cell in urban macro-URLLC, assuming 20 bytes of application data and relevant protocol overhead.

2.3.13 Mobility

The 5G mobility requirement refers to the maximum mobile station speed in such a way that the minimum QoS requirement is still fulfilled. There is a total of four mobility classes defined in 5G: (i) stationary with $0\,km/h$ speed; (ii) pedestrian with $0–10\,km/h$; (iii) Vehicular with $10–120\,km/h$; (iv) high-speed vehicular with speeds of $120–500\,km/h$. The applicable test environments for the mobility requirement are indoor hotspot eMBB (stationary, pedestrian), dense urban eMBB (stationary, pedestrian, and vehicular $0–30\,km/h$), and rural eMBB (pedestrian, vehicular, high-speed vehicular).

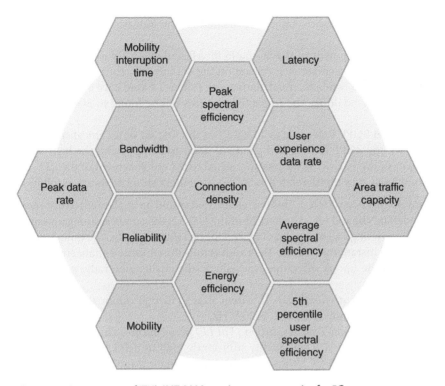

Figure 2.3 As summary of ITU's IMT-2020 requirement categories for 5G.

2.3.14   Mobility Interruption Time

The mobility interruption time refers to the duration of the interruption in the reception between the user equipment (UE) and base station, including RAN procedure execution, radio resource control (RRC) signaling or any other messaging. This requirement is valid for eMBB and URLLC use cases and is set to 0 ms.

2.3.15   Bandwidth

The bandwidth in 5G refers to the maximum aggregated system bandwidth and can consist of one or more radio frequency (RF) carriers. The minimum supported bandwidth requirement is set to 100 MHz, and the RIT/SRIT shall support bandwidths up to 1 GHz for high-frequency bands such as 6 GHz. Furthermore, the ray tracing (RT)/SRIT shall support scalable bandwidth (Figure 2.3).

2.4   The Technical Specifications of 3GPP

2.4.1   Releases

There are two paths within the Release 15 of the 3GPP, one defining the ITU-R – compliant 5G whilst the other route continues developing the LTE under further stage of the LTE-Advanced Pro. The overall 5G as defined via the Release 15 of the 3GPP is described in [10].

Remarkably, the development of the 5G radio interface divides NR into two phases. The first phase is an intermediate step relying on the 4G infrastructure, referred to as non-standalone scenario, while the final, native, and fully 5G-based core and radio system is referred to as standalone. The respective features can be found in 3GPP Release summary [11].

In the path toward 3GPP Release area 15, which is the first set of technical specifications defining 5G, the 3GPP has studied plenty of items to comply with the high-level 5G requirements of the ITU-R. The items are related to the enhancement of the LTE, as well as the completely new topics related to 5G. Some of these study items are as follows:

- *Uplink data compression (UDC).* This is a feature that would be implemented between evolved NodeB (eNB) and UE. The benefit of the UDC is the compression gain especially for web browsing and text uploading, and the benefits cover also the online video performance and instant messaging. The result is thus the capacity enhancement and better latency in the uplink direction especially in the challenging radio conditions [12].
- *Enhanced LTE bandwidth flexibility.* This item aims to further optimize the spectral efficiency within the bandwidths of 1.4–20 MHz.
- *Enhanced Voice over Long-Term Evolution (VoLTE) performance.* This item refers to the maintenance of the voice call quality sufficiently high to delay the Single Radio Voice Call Continuity (SRVCC), thus optimizing the signaling and network resource utilization.
- *Virtual reality.* This item refers to the innovation of potential use cases and their respective technical requirements on virtual reality.
- *NR-based access to unlicensed spectrum.* This item is related to the advanced phase of the Release 15 by investigating further the feasibility of license assisted access (LAA) both below and above 6 GHz frequency band, such as 5, 37, and 60 GHz.
- *NR support on nonterrestrial networks.* This item refers to the channel models and system parameters respective to the nonterrestrial networks, to support satellite systems as one part of the 5G ecosystem.
- *Enablers for network automation for 5G.* This item relates to the automatic slicing network analysis, based on network data analytics (NWDA).
- *System and functional aspects of energy efficiency in 5G networks.* This item refers to the overall topic of energy efficiency (EE). The subtopics include EE key performance indicators (KPIs) relevant to the 5G system, including the ones identified by European Telecommunications Standards Institute (ETSI) TC EE, ITU-T (Telecommunication Standardization Sector of ITU) SG5, and ETSI Network Functions Virtualization (NFV) ISG (Industry Specification Group), as well as the feasibility of the operation and maintenance to support the 5G EE and innovation in general for better optimizing the energy optimization.

2.4.2 Security Requirements for 5G

The 3GPP has defined a set of security requirements for 5G in technical specification TS 33.501. Figure 2.4 summarizes the key elements and functionalities, and the following sections detail the respective key requirements.

The respective security requirements are summarized in the following sections.

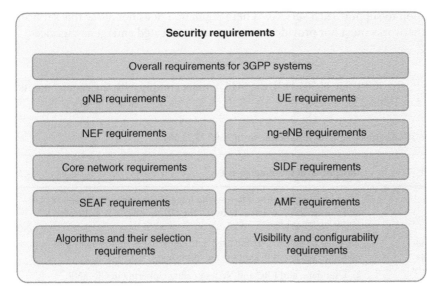

Figure 2.4 The 3GPP security requirement categories.

2.4.2.1 Overall Security Requirements

The generic security requirements of the 3GPP TS 33.501 cover the following cases:

- UE must include protection against bidding-down attack.
- Network must support the subscription authentication and key agreement (AKA).

The bidding-down attack refers to the intentions to modify the functionality of the UE and/or network to make it believe that the networks and/or UE would not support security features. This may lead to security breaches by forcing reduction of the security measures between the UE and network.

The subscription AKA is based on a 5G-specific identifier SUPI (Subscription Permanent Identifier). There is also a protected network identifier for the mutual authentication procedure so that the UE can reliably authenticate the network. The serving network authentication refers to the procedure where the UE authenticates the serving network identifier based on successful key use as a result of the AKA.

Furthermore, the requirement for the UE authorization refers to the procedure where the serving network authorizes the UE through authenticated SUPI and the subsequent subscription profile, which is provided by the home network.

The serving network authorization, on the other hand, refers to the procedure where the home network authorizes the serving network and the UE can rely on the authenticity of the connected serving network.

Lastly, the access network authorization refers to the procedure where UE can rely on all the access networks after they are authorized by the serving network.

Each mobile equipment (ME) must support unauthenticated emergency services, which refers to the possibility to make an emergency call even without the subscription credentials of the Universal Integrated Circuit Card (UICC). In other words, the ME must be able to execute emergency calls without the Universal Subscriber Identity Module (USIM). This requirement applies only to serving networks that are dictated by local regulation to support such service. Where regulation does not allow this service, the serving networks must not provide with the unauthenticated emergency services.

2.4.2.2 UE

As dictated by the 3GPP TS 33.501 [13], the 5G UE must comply with the requirements summarized in Table 2.1.

Table 2.1 The 5G UE requirements as interpreted from 3GPP TS 33.501.

Requirement	Description of the requirement
Ciphering and *integrity protection*	Must be equal for the ciphering and integrity protection between the UE and the ng-eNB, as well as between the UE and the eNB as dictated in 3GPP TS 33.401.
User data and signaling data *confidentiality*	UE must be able to activate used data ciphering as indicated by gNB. In addition, UE needs to support the following: 1. User data, and RRC and NAS signaling ciphering in UE–gNB. 2. NEA0, 128-NEA1, and 128-NEA2 ciphering algorithms, and optionally 128-NEA3. 3. Ciphering algorithms detailed in the 3GPP 33.401 if the UE can provide E-UTRA connectivity to 5GC. UE should use confidentiality protection when regulations so permit, although the confidentiality protection of user data between the UE and the gNB, and confidentiality protection of RRC-signaling and NAS-signaling, are left optional.
User data and signaling data *integrity*	The UE supports the following: 1. Integrity protection and replay protection of user data in UE–gNB. 2. Integrity protection and replay protection of RRC and NAS-signaling, and integrity protection of user data as dictated by the gNB. 3. NIA0, 128-NIA1, 128-NIA2 integrity protection algorithms. The UE may implement optionally also the 128-NIA3. 4. Integrity algorithms as detailed in 3GPP TS 33.401 if the UE supports E-UTRA connectivity to 5G core network. Nevertheless, the actual use of user data integrity protection in UE–gNB interface is left optional, as it adds overhead and processing load. 5. Integrity protection of the RRC and NAS signaling with some exceptions indicated in 3GPP TS 24.501 and TS 38.331.
Secure *storage* and *processing* of subscription credentials	For the storage and processing of the subscription credentials to access the 5G network, the following applies: 1. The subscription credentials are integrity-protected within the UE using a tamper-resistant secure hardware component. 2. The long-term K keys of the subscription credentials are confidentiality-protected within the UE using a tamper-resistant secure hardware component. 3. The long-term keys of the subscription credentials are never presented in the clear format outside of the tamper-resistant secure hardware component. 4. The authentication algorithms for the subscription credentials are executed within the tamper-resistant secure hardware component. 5. A security evaluation and assessment must be possible to assess the compliance of security requirements of the tamper-resistant secure hardware component.

Table 2.1 (Continued)

Requirement	Description of the requirement
Subscriber *privacy*	As for the subscriber privacy, the following applies: 1. The UE supports 5G-GUTI. 2. The SUPI is not transferred in clear text format via 5G RAN. The exception of this rule is the routing information (Mobile Country Code, MCC and Mobile Network Code, MNC), which can be exposed. Please note that SUPI privacy protection is not required for unauthenticated emergency call. 3. USIM houses the home network public key. 4. ME supports the null-scheme. This applies in the scenario when the home network has not provisioned the public key within USIM, and there is thus no SUPI protection in initial registration procedure, and the ME uses null-scheme. 5. Either USIM or ME can calculate the SUCI upon MNO rules, which USIM indicates. 6. Home network operator controls the provisioning and updating of the home network public key within the tamper-resistant hardware, as well as the subscriber privacy enablement. For the user and signaling data integrity, the following applies: 1. The UE supports integrity protection and replays protection of user data and RRC/NAS signaling in UE–gNB interface. 2. gNB indicates the use and UE activates accordingly integrity protection of user data. 3. Support of NIA0, 128-NIA1, and 128-NIA2 integrity protection algorithms, and optionally 128-NIA3. Nevertheless, the actual use of integrity protection of the user data in UE–gNB is optional. 4. The UE houses the integrity algorithms as defined in 3GPP TS 33.401, if the UE supports E-UTRA connectivity to 5GC. 5. Integrity protection of the RRC and NAS-signaling is mandatory to use, except in cases summarized in 3GPP TS 24.501 and TS 38.331.

2.4.2.3 gNB

As dictated by the 3GPP TS 33.501 [13], the 5G based station, i.e. fifth-generation NodeB (gNB), must comply with the requirements summarized in Table 2.2.

2.4.2.4 ng-eNB

For the detailed security requirements of ng-eNB, please refer to the security requirements as specified for gNB above and in the 3GPP TS 33.401.

2.4.2.5 AMF

The 3GPP TS 33.501 includes requirements for the AMF as for signaling data confidentiality and integrity, and subscriber privacy.

The AMF supports ciphering of non-access stratum (NAS)-signaling by applying one of the New Radio Encryption Algorithms (NEA) from the set of NEA0, 128-NEA1, and 128-NEA2. The AMF may also support the optional 128-NEA3. Also, the confidentiality protection of the NAS-signaling is optional to use, although it is encouraged to be applied when regulations so permit.

Table 2.2 Key requirements for gNB.

Requirement	Description of the requirement
User data and signaling data *confidentiality*	For the user data and signaling data confidentiality, the following applies: 1. The gNB supports ciphering of user data and RRC-signaling in UE–gNB. 2. Session management function (SMF) dictates the use, and the gNB activates ciphering of user data. 3. The gNB supports NEA0, 128-NEA1, and 128-NEA2 ciphering algorithms, and optionally, the 128-NEA3. 4. It is optional to support confidentiality protection of user data and RRC-signaling in UE–gNB. It should be noted that confidentiality protection is encouraged to be applied when regulations so permit.
User data and signaling data *integrity*	For the user and signaling data integrity, the following applies: 1. The gNB supports integrity protection and replays protection of user data and RRC signaling in UE–gNB. 2. SMF dictates and gNB activates integrity protection of user data. 3. The gNB supports NIA0, 128-NIA1, and 128-NIA2 integrity protection algorithms, and optionally 128-NIA3. Nevertheless, the actual use of integrity protection of the user data in UE–gNB is left as optional. 4. RRC signaling messages excluding the cases detailed in the 3GPP TS 38.331 are integrity-protected. The enablement of NIA0 in gNB depends on the regulatory requirements for the support of unauthenticated emergency session.
gNB setup and configuration	For the gNB setup and configuration, the following applies: 1. When O&M sets up and configures gNBs, if MNO so dictates, the gNB authenticates and authorizes the procedure according to the certification scenario detailed in 3GPP TS 33.310. This is to prevent external parties to modify gNB settings and configuration by using local or remote access. 2. The communication in O&M–gNB is confidentiality, integrity and replay protected. The security associations between O&M, gNB, and the 5G core network are detailed in the 3GPP TS 33.210 and TS 33.310. 3. The gNB can authorize the intentions for (authorized) software and data changes. 4. Secure environment is applied for execution of sensitive parts of the boot-up. 5. Confidentiality and integrity of software, which is transferred to gNB, needs to be ensured, and the software update of gNB must be verified upon installation as dictated in 3GPP TS 33.117.
gNB key management	For the key management in the gNB, the following applies: 1. The key protection is of utmost importance when 5G core network provides the keying material for the gNBs. 2. The parts of gNB deployment storing or processing keys in clear format must be protected from physical attacks, possibly applying secure physical environment. Keys in such secure environment must be stored only there, excluding cases allowed by 3GPP specifications.
Handling of user and control data for the gNB	For handling gNB user and control plane data, the following applies: 1. Parts of gNB storing or processing user or control plane data in clear format must be protected against physical attacks. Otherwise, the entity needs to be placed in a physically secure location for storing and processing the user and control plane data in clear format.

Table 2.2 (Continued)

Requirement	Description of the requirement
Secure environment of the gNB	The secure environment protects sensitive information and operations from unauthorized access or exposure. For the secure environment logically defined within gNB, the following applies:
	1. It supports secure storage for sensitive data and execution of sensitive functions such as encryption and decryption of user data. 2. It also supports the execution of sensitive parts of the boot-process. 3. The integrity of the secure environment must be assured. 4. The access to the secure environment must be authorized.
gNB *F1* interfaces	The set of NBs with split DU-CU (distributed unit, centralized unit) implementations based on *F1* interface is defined in 3GPP TS 38.470. For the gNB *F1* interface, the following applies:
	1. Signaling traffic F1-C as defined in 3GPP TS 38.470, F1-C signaling bearer as defined in TS 38.472, and user plane data can use *F1* interface in DU–CU. 2. gNB supports confidentiality, integrity, and replay protection for F1-C signaling bearer, and the management traffic on F1-C interface (as per 3GPP TS 38.470) must be integrity, confidentiality, and replay protected. 3. The gNB supports confidentiality, integrity, and replay protection on the gNB DU–CU F1-U interface for user plane.
gNB E1 interfaces	The 3GPP TR 38.806 describes the *E1* interface. For the gNB *E1* interface, the following applies:
	1. The 3GPP TS 38.460 defines the principles for the gNBs with split DU–CU implementation, including open interface between CU–CP and CU–UP using the *E1* interface. 2. The E1 interface between CU–CP and CU–UP is confidentiality, integrity, and replay protected.

The AMF supports integrity protection and replay protection of NAS-signaling via one of the following New Radio Integrity protection Algorithms (NIA): NIA-0, 128-NIA1, and 128-NIA2, or via optional 128-NIA3. It should be noted that AMF must disable the NIA-0 in those deployments where support of unauthenticated emergency session is not a regulatory requirement. The NAS signaling messages must be integrity-protected with exceptions summarized in the 3GPP TS 24.501.

The AMF supports the triggering of the primary authentication based on the Subscriber Concealed Identifier (SUCI). The AMF also supports the procedures to assign and reallocate 5G-GUTI to the UE. The AMF can confirm SUPI derived from UE and home network and may only proceed with the service to the UE if the confirmation is successful.

2.4.2.6 SEAF

The key security requirement for the SEAF is to support primary authentication using SUCI.

2.4.2.7 NEF

The Network Exposure Function (NEF) provides external exposure of network function (NF) capabilities to the application function (AF). AFs thus interact with the selected

NFs via the NEF of the 5G network. To do so, the NEF must be able to determine if the AF is authorized to interact with the NFs. The communication in NEF–AF supports integrity protection, replay protection, and confidentiality protection. The interface also supports mutual authentication.

There are restrictions for the handled data. Thus, internal 5G core information, e.g. Data Network Name (DNN) and single network slice selection assistance information (S-NSSAI), must not be sent outside the 3GPP mobile network operator (MNO) network. Furthermore, NEF must not send SUPI outside of the 3GPP operator domain.

2.4.2.8 SIDF

The task of Subscription Identifier Deconcealing Function (SIDF) is to resolve the SUPI from the SUCI. This happens based on the protection scheme meant for generating the SUCI. The SIDF is in the home network, and in practice, unified data management (UDM) offers the SIDF services.

2.4.2.9 Core Network

The 3GPP TS 33.501 collects security requirements for the 5G core network. These include trust boundaries, service-based architecture, and end-to-end interconnectivity. The following summarizes the respective key statements.

Trust Zones The 5G systems provide the possibility for MNO to divide the network into trust boundaries, or zones. The default assumption is that two different MNOs are not sharing a single trust zone. The data transferred between the trust zones is handled by the principles of the service-based architecture of the 5G, although it also can be protected by applying end-to-end security based on Network Domain Security (NDS) and IP, as described in 3GPP TS 33.210.

Service-Based Architecture The 5G system allows the service-based architectural model, which is novel concept compared to the previous generations. The respective security requirements for service registration, discovery, and authorization include the following:

- NF service-based discovery and registration supports confidentiality, integrity, and replay protection, and there must be assurance that the NF discovery and registration requests are authorized.
- NF service-based discovery and registration can hide the topology of the NFs between trust domains, e.g. between home and visited networks.
- NF service request and response procedure supports mutual authentication between NFs.
- There must be a mutual authentication between Network Repository Function (NRF) and the set of NFs requesting service from it.

End-to-End Core Network Interconnection Security The end-to-end security solution supports application-layer mechanisms for addition, deletion, and modification of message elements handled by intermediate nodes, excluding special message elements such as those related to routing in Internet Protocol Packet eXchange (IPX) environment. These exceptions are detailed in 3GPP TS 33.501. By default, the solution provides end-to-end confidentiality, integrity, and authenticity between source and destination network by utilizing the Security Edge Protection Proxy (SEPP) concept. It is required that the solution would have only minimal impact on functionality, performance, interfaces, and equipment compared to 3GPP network elements, relying on standard security protocols.

The role of the SEPP is to function as a nontransparent proxy node. Its tasks include the protection of application layer control plane *N32* interface-based messages between two NFs of separate Public Land Mobile Networks (PLMNs), mutual authentication, and ciphering procedures with the counterparty SEPP, key management for securing messages on the *N32* interface between two SEPPs, and topology hiding. Furthermore, the SEPP functions as a reverse proxy providing a single point of access and control to internal NFs.

In the *N32* interface, the integrity protection applies to all of the transferred attributes. *N32* also applies confidentiality protection for the authentication vectors, cryptographic material, location data including Cell ID and SUPI.

2.4.2.10 5G Algorithms

The initial authentication procedure of the 5G is still based on the same AKA concept as in previous generation. Thus, the same respective algorithms, i.e. MILENAGE and TUAK, are still valid. Nevertheless, the rest of the security as for the key derivation is renewed in 5G. 5G brings thus along with new algorithms for the integrity, confidentiality, and ciphering.

Table 2.3 summarizes the 5G encryption (ciphering) algorithms and Table 2.4 summarizes the integrity protection algorithms [13].

The UE and serving network will negotiate the utilized algorithm based on the MNO policies and equipment security capabilities. This applies to the ciphering and integrity protection of RRC signaling and user plane for UE–gNB, and ciphering and integrity protection of RRC signaling and ciphering of user plane for UE–ng-NB. The algorithms also take place in NAS ciphering and NAS integrity protection for UE–AMF.

Table 2.3 5G ciphering algorithms.

Algorithm	Description
NEA0	Null ciphering algorithm
128-NEA1	128-bit SNOW 3G-based algorithm as referenced in 3GPP TS 35.215
128-NEA2	128-bit AES-based algorithm in CTR mode
128-NEA3	128-bit ZUC-based algorithm as referenced in 3GPP TS 35.221

Table 2.4 5G integrity protection algorithms.

Algorithm	Description
NIA0	Null integrity protection algorithm
128-NIA1	Null integrity protection algorithm-based on SNOW 3G (3GPP TS 35.215)
128-NIA2	128-bit AES-based algorithm in CMAC mode
128-NIA3	128-bit ZUC-based algorithm as referenced in 3GPP TS 35.221

The security capabilities of a multiple Radio Access Technology (RAT) UE-supporting Evolved UMTS Terrestrial Radio Access Network (E-UTRAN) and NR connected to the 5G core network include the LTE and NR algorithms. The UE security capabilities are protected against so-called bidding-down attacks referring to an attacker forcing the connectivity to be directed via older and less protected mobile generation networks.

2.5 NGMN

The NGMNs is an initiative that also contributes, among many other technology areas, to the 5G development.

The 5G vision of NGMN is as follows:

> 5G is an end-to-end ecosystem to enable a fully mobile and connected society. It empowers value creation toward customers and partners, through existing and emerging use cases, delivered with consistent experience, and enabled by sustainable business models.

To comply with these aspects, Ref. [14] lists the following items that the NGMN Alliance considers priorities:

- UX
- System performance
- Device
- Enhanced services
- New business models
- Network deployment, operation, and management

2.5.1 User Experience

UX refers to the end-UX while consuming a single or multiple simultaneous service. In the ever-increasing competitor environment, this is one of the key aspects that differentiates the stakeholders and can either fortify or break the customer relationship.

Especially for the 5G era, the enhanced networks need to deliver smooth and consistent UX independent of the time or location. The KPIs contributing to the UX include the minimum needed data rate and maximum allowed latency within the service area, among a set of various other parameters that are configurable, including their allowed ranges of variations, by each MNO. These service-dependent KPIs and their respective values are investigated by many stakeholders at present, including ITU-R, which has designed the minimum requirements for the 5G systems, so they may be called ITU-compliant 5G (IMT-2020) systems.

The UX data rate is dictated by the vision of providing broadband mobile access everywhere. In terms of use cases of NGMN, the practical requirement is to provide at least 50 Mb/s peak data rate for a single user over the whole service area of the 5G. The NGMN emphasizes that this value must be provided constantly over the planned area, including the cell edge regions.

The UX mobility requirements of the NGMN are use cases. They include the following:

- *High-speed train.* The train's velocity could be over 500 km/s. The 5G must support the data transmission during the train trips, including HD video streaming and video conferencing.
- *Remote computing.* This category includes both stationary (e.g. home office) and mobile environments (e.g. public transportation).
- *Moving hot spots.* This category refers to the extension of the static hot-spot concept to better tackle with the high mobility demands in 5G era. As dictated by the NGMN, 5G shall complement the stationary mode of planning of capacity by adding the nonstationary, dynamic and real-time provision of capacity.
- *3D connectivity.* This is important for, e.g. aircrafts, for providing advanced passenger services during the flight comparable with the UX on the ground.

Other important 5G requirements, per NGMN Alliance, that are related to the UX are the system performance in general, connection density, traffic density, spectrum efficiency, radio coverage area, resource and signaling efficiency, and system performance.

In addition to these requirements for the broadband access everywhere and high-user mobility, the NGMN has formed an extensive list of other cases and their requirements for 5G. These include the following aspects:

- *Massive Internet of Things (IoT).* The vision beyond 2020 includes remarkably the need to support of a considerably higher number of all types of devices compared to current markets, such as smart wearables and clothes equipped with multitude of sensors, sensor networks, and mobile video surveillance applications.
- *Extreme real-time communications.* These cases represent the environment for the most demanding real-time interactions, and for the full compliance may require more than one attributes amongst extremely high data throughput, high mobility, and critical reliability. A real-world example in this category includes the tactile Internet, which requires tactile controlling and audiovisual feedback. Robotic control falls into this category.
- *Lifeline communication, such as disaster relief and emergency prediction.* Along with the 5G, there will be increasing expectations to rely on the mobile communications in the role of a lifeline in all the situations and in wide areas. Remarkably, 5G would need to be optimized for robust communications in case of natural disasters such as earthquakes, tsunamis, floods, and hurricanes.
- *Ultrareliable communications.* This category includes critical cases such as automated traffic control and driving, i.e. self-driving vehicles, collaborative robots and their respective networking, life-critical health care, and people's emergency services such as remote surgery, 3D-connectivity for drones (large coverage for ensuring packet transport), and public safety (real-time video from emergency areas).
- *Broadcast-like services.* This category extends the traditional model of information broadcast such as TV networks by adding means for more advanced interaction.

2.5.2 Device Requirements

The role of the ever-evolving and increasingly complex devices is important in the 5G era. The HW, SW, and operating system (OS) continue to evolve, and the assurance of the correct and fluent functioning of those is of utmost importance. Not only do the devices such as isolated and individual elements need to perform correctly, but they

Table 2.5 NGMN Alliance's device requirements.

Requirement	Description
MNO controllability	Ability for NW or UE to choose desired profile depending on the QoS need, element capabilities and radio conditions; Ability for MNO to manage the HW and SW diagnostics, fixes, and updates; Ability to retrieve performance data from the UE as a basis for further optimization and customer care.
Multi-band and multi-mode	A sufficiently wide support of RF bands and modes (Time Division Duplex [TDD], Frequency Division Duplex [FDD], and mixed) is needed for efficient roaming scenarios. Ability to take advantage of simultaneous support of multiple bands optimizes the performance. Aggregation of data from different RATs and carriers is required.
Power efficiency	Whether the issue is support for a vast amount of IoT devices or devices that are in remote areas, enhanced battery efficiency is important. For the consumer devices (i.e. smart devices), the minimum battery life requirement is 3 days, while autonomously working IoT devices need to function up to 15 years.
Resource and signal efficiency	The 5G devices require optimized signaling as one of the ways to provide long battery life.

also are used increasingly as relay elements with the other 5G devices, especially in the IoT environment, as well as in various consumer use cases.

Table 2.5 summarizes the high-level device requirements as informed by NGMN Alliance.

2.5.3 Enhanced Services

There are various special requirements for enhanced functionality of 5G, related to connectivity, location, security, resilience, high availability, and reliability. The connectivity is related to the transparency, which is a key for delivering consistent experience in such a heterogeneous environment 5G will represent.

It is estimated that 5G combines both the native 5G networks as well as legacy Radio Access Technologies, especially as the 3GPP LTE system is evolving in a parallel fashion with the 5G system. Furthermore, 5G allows UE to connect simultaneously to more than one RAT at the time, including carrier aggregated connections. It should be noted that this type of connectivity may also involve other systems such as IEEE 802.11ax, which is state-of-the-art, high-efficiency Wi-Fi.

The 5G systems automatically optimize the respective combinations based on the most adequate achievable UX, which requires a fluent and seamless transition between the RATs. Detailing further, both the Inter-RAT and Intra-RAT mobility interruption time for all the RAT types and technologies need to be unnoticeable. Also, the inter-system authentication needs to be seamless, including the set of both internal 3GPP as well as non-3GPP RATs.

Another enhanced service requirement is the location, which is a key for contextual attributes. More concretely, the network-based positioning in 5G should be able to achieve accuracy of 1–10 m at least 80% of the occasions, and it should be better than 1 m indoors. Driven by high-speed devices, this accuracy must be provided in real time.

Furthermore, the network-based location of 5G should be able to cooperate with other location technologies. One example of such cooperation is the data delivery between partner sources so that the set of complete information enhances the final accuracy of the location. The NGMN also puts requirement on the cost of the 5G location-based solution, which should not exceed the currently available partner options based on, e.g. satellite systems or 4G solutions.

The NGMN also has requirements for the security of the 5G. It will support a variety of diverse applications and environments, including both human and machine-based communications. This means that there will be a greatly increased amount of sensitive data transferred over 5G networks, and it will need enhanced protection mechanisms beyond the traditional models of protecting the communications between nodes and end-to-end chain. This enhancement aims to better protect user data, to create new business models, and protect the users and systems against cybersecurity attacks.

5G will support a wide range of applications and environments, from human-based to machine-based communication, and it will deal with a huge amount of sensitive data that need to be protected against unauthorized access, use, disruption, modification, inspection, attack, etc. Since 5G offers services for critical sectors such as public safety and eHealth, the importance of providing a comprehensive set of features guaranteeing a high level of security is a core requirement for 5G systems.

2.5.4 The 5G Security

It can be generalized that the trust of the 5G system is a concept consisting of security, identity, and privacy:

- Specifically, for the security aspects, the NGMN states that the operator is the partner for state-of-the-art data security, running systems that are hardened according to recognized security practices, to provide security levels for all communication, connectivity, and cloud storage purposes.
- The identity refers to the operators' special role in the security, which is the trusted partner for master identity. The MNO provides secure, easy-to-use single sign-on and user profile management to fit all communication and interaction demand.
- The privacy refers to the operators' special role in partnering to safeguard sensitive data, while ensuring its transparency.

The traditional role of MNOs has been to provide subscriber authentication. As has been the case with the 4G stage, the 5G authentication is based on a robust platform that can be used as a base for MNOs to deploy single-sign-on services. A logical role for the 5G MNO is to act as an identity provider for the networks' own services as well as for external entities dealing with the MNOs' customers. MNO can thus provide transparent identification and seamless authentication to application services on behalf of the user.

What is then the role of the traditional subscriber identity module (SIM)/UICC that has been used for the subscriber identity vault since the very first commercial deployment of the 2G networks? There have been intensive discussions on the subject in standardization of the security of the mobile communications, including ETSI smart card platform (SCP). The consensus has been recently reached, and the outcome is that the secure element (SE) for storing the subscriber data does not need to rely any more on standalone SIM/UICC element. Instead, if the secure storing of this subscriber data combined with the secret data that allows the access to the mobile networks can be ensured in such a way that no external parties can intercept nor modify it, any secured physical entity is allowed for such storing.

The UICC is still valid for such storing, but equally, the storage can be, e.g. external nonvolatile memory of the mobile device while the handling of the cryptographic functions and execution of the plain data can happen, e.g. within a SoC (system on chip). There may be several technical terms for this concept, some being iUICC (Integrated Universal Integrated Circuit Card) and soft-SIM. In any case, the data used to access operator network remain solely owned by the operator running this network.

Interpreting further the security needs and requirements, the NGMN states that the system shall offer the capability to protect 5G customers from common security threats. These threats may include impersonation and traffic eavesdropping. This means that there is a need to increase the level of trust that is associated with the network subscribers' identity. In addition, the spirit of the 5G security requirements is that the planning of the security solutions for the operational phase of the networks, e.g. when the handover takes place between different RANs, must be better that in previous mobile communication generations, though in such a way that the efficiency to perform such protection mechanisms is higher.

Another trend is that 5G needs to enhance the current solutions for the privacy of the users as well as the equipment in the machine-type communications. In addition to the identity itself, the aspects requiring special attention include the information revealing the set of subscribed services, location and presence, mobility patterns, network usage behavior, and utilization of applications.

The adequate protection of the radio interface continues to be crucial, but also the higher-level protection needs to be reviewed in 5G. In fact, 5G is designed in a bearer-independent way in such a manner that the additional security and protection mechanisms may create additional value for advanced business, e.g. for extending the network protection into the Internet and end-to-end chain of the machine communications. This could better tackle the increasing threat of interoperator fraud and illegal use of international signaling networks. The 5G can handle this threat in 5G roaming, as the respective signaling protocols can enable the home network to verify that a user is attached to a serving network that claims it is. This aspect is thus the same as in domestic cases; one of the principals and priority tasks of the involved parties is to ensure that the one (consumer or machine) communication in the network is the one it/he/she claims to be.

The potential network vulnerabilities can be estimated to increase along with the huge increase of the number of the simultaneously communication parties in 5G, as the number of respective IP protocols for both control and user planes over such an overwhelming NFs increases. One of the very special new aspects is the introduction of extremely low-cost machines and open-OS smartphones, which by default open very easily the ports for mobile malware. These dangers can extend, not only between the devices and services, but also within the infrastructure of the 5G radio and core networks. As the NGMN states, there is thus an urgent need to fortify the following aspects of the networks to prepare MNOs to better tackle the new threats (intentional as well as accidental ones) by improving:

- Resilience against signaling-based threats, including direct attacks and overloading of the signaling channels;

- Security design for very low latency use cases for initial signaling and during the communications;
- Security requirements originated in the 4G technical specifications.

The NGMN mentions also the special needs for the public safety and mission-critical communications. The aim of the development is to reduce costs for these modes. As 5G supports the emergency communications in the same or better fashion as before, it also provides basic security functions in emergency situations, even if part of the network infrastructure and its security functions should be destroyed. Under these special conditions, 5G needs to continue providing protection against malicious attacks, including the ability to resist advanced radio jamming attacks over the 5G service area and attempts to compromise small-cell nodes distributed over wide geographical areas.

2.6 Mobile Network Operators

In a parallel fashion with the standardization work of 5G, several MNOs have been testing and developing concepts for 5G. These activities help in better understanding the performance of the NR and core network concepts and assist in the evaluation work of the standardization bodies.

As an example, the US-based MNO Verizon is active in evaluating the 5G concepts and has established a dedicated Verizon 5G Technology Forum, V5GTF, to study the item thoroughly [15]. The Forum is established in cooperation with Cisco, Ericsson, Intel, LG, Nokia, Qualcomm, and Samsung, with the focus on creating a common and extendable platform for Verizon's 28/39 GHz fixed wireless access trials and deployments.

As informed by the V5GTF, the participating partners have collaborated to create the 5G technical specifications for the 5G radio interface of OSI (Open Systems Interconnection model) layers 1, 2, and 3. The specifications also define the interfaces between the UE and the network for ensuring interoperability among network, UE, and chipset manufacturers.

The initial release, which has been ready since 2017, includes the V5G.200 series for the physical layer 1. The V5G.300 series describes layers 2 and 3 for Medium Access Control, Radio Link Control, Packet Data Convergence Protocol, and Radio Resource Control.

Many other MNOs are actively driving 5G development. Prior to the commercial Release-15-based deployments, some of the most active stakeholders were AT&T, Sprint, and DoCoMo.

2.7 Mobile Device Manufacturers

As an example, from the mobile network equipment manufacturing domain, Ericsson has identified the following aspects for the requirements [16]:

- *Massive system capacity*. The remarkable aspect in this category is that to support the augmented capacity, 5G networks must transfer data with lower cost per bit compared to today's networks. Also, a relevant aspect is the energy consumption, as increased data consumption also increases the energy footprint of the networks; thus, 5G needs to optimize the energy better than is done today.

- *High data rates.* This aspect refers to considerably increasing sustainable, real-life data rates in much wider areas compared to previous generations, whereas the focus has traditionally been on the dimensioning of the networks based on the peak data rates. Concretely, 5G needs to support data rates of 10 Gb/s in limited environments such as indoors, whereas several hundreds of Mb/s must be offered in urban and suburban environments. Finally, the idea of 5G is to provide 10 Mb/s data rates almost everywhere.

As a comparison, Nokia, in alignment with merged Alcatel Lucent, summarizes expected 5G use cases and requirements in [17].

5G presents totally new solutions in many domains while it also is largely an evolution path from the previous mobile generations. One of the new aspects is related to IoT, including the M2M (machine-to-machine) communications. This results in a vast, new business, and new stakeholders produce products and services for IoT markets and relying increasingly on 5G infrastructure.

It is worth noting that 5G does not represent IoT as such, but it suits well the related communications, e.g. between machines. As in some examples, it will support a much higher number of simultaneously connected devices, which can be thousands under a single radio cell.

References

1 ITU, "Minimum requirements related to technical performance for IMT-2020 radio interface(s)", ITU, 23 February 2017. https://www.itu.int/md/R15-SG05-C-0040/en. Accessed 1 March 2017.

2 ITU-R, "Rep. ITU-R M.2410-0, Minimum requirements related to technical performance for IMT-2020 radio interface(s)", ITU, 2017 November.

3 ITU, "ITU agrees on key 5G performance requirements for IMT-2020," 23 February 2017. http://www.itu.int/en/mediacentre/Pages/2017-PR04.aspx. Accessed 1 March 2017.

4 NGMN, "NGMN", 04 July 2018 https://www.ngmn.org/home.html. Accessed 4 July 2018.

5 ITU, "ITU towards "IMT for 2020 and beyond," 2016. http://www.itu.int/en/ITU-R/study-groups/rsg5/rwp5d/imt-2020/Pages/default.aspx. Accessed 7 September 2016.

6 ITU-R, "ITU-R Recommendation M.2083", ITU-R, 04 07 2018. http://www.itu.int/rec/R-REC-M.2083. Accessed 04 July 2018.

7 ITU-R, "Itu-R Recommendation M.2376, Technical feasibility of IMT in bands above 6 GHz", ITU, July 2015. https://www.itu.int/pub/R-REP-M.2376. Accessed 4 July 2018.

8 ITU, "Press Release: ITU agrees on key 5G performance requirements for IMT-2020", ITU, 27 February 2017. http://www.itu.int/en/mediacentre/Pages/2017-PR04.aspx. Accessed 4 July 2018.

9 ITU, "ITU-R M.[IMT-2020.EVAL]", ITU, 2017.

10 3GPP, "Release 15", 3GPP, 2018. http://www.3gpp.org/release-15. Accessed 4 July 2018.

11 3GPP, "3GPP Features and Study Items", 3GPP, 2018. http://www.3gpp.org/DynaReport/FeatureListFrameSet.htm. Accessed 4 July 2018.

12 3GPP, "R2-161747, Discussion on uplink data compression", 3GPP, 2018. http://portal.3gpp.org/ngppapp/CreateTdoc.aspx?mode=view&contributionId=687049. [Accessed 4 July 2018].

13 3GPP, "Security architecture and procedures for 5G system, Release 15, V.15.1.0", 3GPP, 2018.

14 NGMN (2015). *NGMN 5G White Paper*. NGMN Alliance.

15 Verizon, "Verizon 5G Technical Forum", Verizon, 2018. http://www.5gtf.net. [Accessed 28 July 2018].

16 Ericsson (2015). *5G Radio Access*. Ericsson AB.

17 Nokia (2014). *5G Use Cases and Requirements*. Espoo: Nokia Networks.

3

Positioning of 5G

3.1 Overview

The evolving mobile communications technologies provide us with vastly increasing data rates, faster response times, and increasingly fluent means to access more content than ever in the history of telecommunications. Along with the development of the standards, the commercial solutions keep appearing for our selection. Often in this highly dynamic environment, the vast amount of technical and marketing terms challenges us to capture the real meaning of the messages that operators, device vendors, and service providers want to give us. One of the most confusing topics in this field is the definition of the "mobile communications generations" so let's have a look at the principles. The summary of the following sections is based on Refs. [1, 2].

3.2 Mobile Generations

Mobile generations are an evolving story. Early mobile networks paved the way for a totally new era liberating users from communicating in fixed locations back in the late 1980s. In fact, there were some much earlier systems deployed, although they have been basically forgotten in the history. As an example, Swedish automatic MTA (Mobile Telephony System Version A) was deployed in 1956 for operative use in Stockholm and Gothenburg. As the equipment was bulky, it was suitable merely for a car-mounted environment. This initiation was still ahead of its time, so it did not have commercial success. As another example from 1971, the technology was finally mature enough in Finland for the commercial launch of ARP (Auto Radio Phone), which functioned in 160 MHz band until 2000 – serving customers especially in remote areas for whole three decades. Parallel systems were also adopted in many locations in 450 MHz band. All these initiations may be referred to as "Generation 0" with relatively small user base and limited portability.

After the pre-1G system, the markets started to take off. Table 3.1 summarizes the main aspects of the 1G–5G, and the following sections describe the background of each one [2, 3].

5G Explained: Security and Deployment of Advanced Mobile Communications, First Edition.
Jyrki T. J. Penttinen.
© 2019 John Wiley & Sons Ltd. Published 2019 by John Wiley & Sons Ltd.

Table 3.1 A summary of the mobile generations.

Generation	Description
1G	Analogue, first completely or almost completely automatic mobile networks that were meant only for voice calls, although accessory-based solutions were possible to adapt for data usage. The initial systems were based on vehicle-mounted user equipment that could also be used as portable devices. The weight was typically several kilograms, and there was a separate auricular. Some examples of this phase of 1G are NMT-450, Netz-C, and AMPS. In the further development of 1G, there were also handheld devices introduced, although the first ones were large and heavy compared to modern devices – those were not meant for pockets. Example of this phase is NMT-900, launched in Nordic countries 1986–1987.
2G	The most important differentiator of 2G was the digital functionality, which provided integration of messaging and data services into the system and devices. Examples of this generation are GSM and IS-95.
3G	The further development of multimedia-capable systems led into the third generation. The main differentiator of this generation is the possibility to use considerably higher data rates. According to the original set of ITU's performance requirements, the early stage LTE belongs still to the 3G phase.
4G	ITU-R has defined a set of principles and performance requirements for the fourth-generation systems. In the initial phase of the compliance review by ITU, there were two systems that complied with the requirements, i.e. the advanced version of LTE (LTE-A, as of Release 10) and WiMax (as of WirelessMAN-Advanced). As the mobile telecommunications markets have been growing heavily, and the competition is tougher than ever, there have also been parallel interpretations of the 4G capabilities. Often, the Release 8 LTE is interpreted as belonging to 4G, and HSPA+ is considered by various operators to be a 4G system.
5G	Ideas beyond IMT-Advanced provide much higher data rates, with focus on an expected deployment in 2020.

3.2.1 1G

The evolution started to really take off during the 1980s along with the first-generation mobile communications networks that had fully automatic functionalities for establishing voice calls.

Some examples of the 1G networks were AMPS (Advanced Mobile Phone System) in North America and TACS (Total Access Communications System) in the United Kingdom. There also were a diversity of similar kind of networks in other locations of the globe, of which the slightly more international variant was NMT 450 (Nordic Mobile Telephony) and its enhanced version NMT 900 in Nordic countries.

Now obsolete, the 1G cellular networks showed concretely the benefits of location- and time-independent communications, which has resulted in dominating position of mobile users compared to land-line subscriptions from years ago. Looking back at history, the growing popularity of the voice service – thanks especially to the 1G systems – often came as a complete surprise to operators and regulators in the late 1980s and early 1990s, which caused challenges in keeping up with the capacity demand. The more accessible prices of the devices and their utilization triggered the sudden gain in interest, which, in turn, made mobile business grow faster than any estimation could

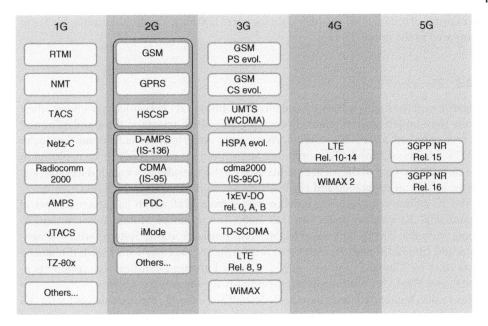

Figure 3.1 Snapshot of some of the most popular commercial mobile communications systems representing different generations.

have indicated at that time. The mobile user base still keeps increasing, and thanks to the growing popularity of multiple subscriptions per customer, we are still far away from the saturation levels, as can be seen from the development of new systems (Figure 3.1). Furthermore, the growing business for Internet of Things (IoT) will contribute to the total growth figures as the automatized communications between machines requires ever-increasing number of subscriptions.

3.2.2 2G

The offered capacity of the 1G systems was rather limited, and regardless of some international functionality especially via NMT, the roaming concept was still fairly new and limited. These were some of the reasons for the further development of fully digitalized, second-generation mobile communication systems like code division multiple access, CDMA-based IS-95 in North America. As for the number of users, GSM (Global System for Mobile Communications) has been the most popular 2G variant to date. GSM also introduced the separation of the terminal and subscriber data storage (SIM, subscriber identity module) for easier changing of the device yet maintaining the mobile telephone number. The 2G systems started to include a growing set of enhanced services, from which the short message service (SMS) is an example of the most successful mobile communications solutions of all times.

The digital 2G systems contained built-in capabilities for data transmission. After further specification efforts, the mobile data services started to take off in the early 2000s, along with the packet switched data transmission. This can be interpreted as the starting point for the development of mobile communications toward the all-IP environment. As an example of today, GSM can offer data speeds of 500 kb/s – 2 Mb/s, considerably faster compared to the initial 9.6 kb/s.

3.2.3 3G

Regardless of the highly advanced and novelty solutions 2G networks have offered, their limits were understood in relatively early phases, which triggered the standardization of even more advanced 3G systems. As a result, American cdma2000 and European UMTS (Universal Mobile Telecommunications System) were ready to be deployed in the early 2000s. One of the main benefits of 3G is the considerably higher data rates, e.g. via 1x-EV-DV of cdma2000 and HSPA (High-Speed Packet Access) of UMTS.

The 3G system requirement guideline comes from ITU, which has accepted a set of commercial 3G systems under the IMT-2000 (International Mobile Telecommunications) umbrella. UMTS (Universal Mobile Telecommunications System), with few variants, and cdma2000 comply with these requirements, together with EDGE (Enhanced Data rates for Global Evolution), DECT (Digital Enhanced Cordless Telecommunications), and WiMAX (Worldwide Interoperability for Microwave Access).

Ever since, higher data rates have been desired and offered. The main driver for the 3G systems was in fact the need of higher data rates as base for the fluent consumption of increasingly advanced multimedia services. Since the early days of 3G networks, mobile data services have developed drastically from the original, some hundreds of kb/s speeds up to current tens of Mb/s category.

Even if it is technically doable to enhance further the older technologies like CDMA and GSM, there is a limit as for the techno-economic feasibility, which justifies the introduction of modernized technologies within the 3G path. The basic reason for this is the better spectral efficiency of the post-2G systems which is a result of the advances in the overall telecommunications technologies, processor performance and due to better understanding of the theories in the constant efforts to reach the performance limits. This is thus a logical moment for continuing the deployment of enhanced 3G systems of which Long-Term Evolution (LTE) is the most popular representative and was first available around 2010–2011. It offered better spectral efficiency than any of the previous systems have been able to do and moved the data service reference to new decade by providing hundreds of Mb/s data rates. WiMAX is another representative in the advanced 3G path, although its wider commercial use is yet to be seen.

3.2.4 4G

Up to 3G era, the terminology for distinguishing between the mobile system generations was mostly understandable and the commonly agreed definition of 3G, based on ITU's IMT-2000 requirement umbrella, is logically aligned with the previous 2G and 1G networks. Nevertheless, the term "4G" seems to be somewhat confusing. There are several interpretations, the strictest one coming from the ITU that accepts only the evolved versions of LTE and WiMAX to the 4G category. Logically thinking, the basic version of LTE that is defined in the third-generation partnership project (3GPP) Release 8 specifications is relatively close to 4G and could thus be considered as a "beyond 3G, pre-4G" system, also sometimes referred to as 3.9G technology in nonstandard communications.

In practice, many operators and device manufacturers are interpreting early-stage LTE to belong to fourth generation as such.

Investigating the history behind the definition of 4G, ITU's radiocommunication sector (ITU-R) completed the assessment of six candidate submissions for the global 4G mobile wireless broadband technology by 21 October 2010. According to the ITU-R terminology, the fourth generation refers to IMT-Advanced that contains various technical requirements, e.g. for the data rates. The proposals resulted in two 4G technologies, "LTE-Advanced" and "Wireless MAN-Advanced," i.e. IEEE 802.16m, which also is commonly known as "WiMAX 2." Only these solutions are currently, officially recognized by ITU-R as 4G technologies, although the practical interpretation of the mobile telecommunications industry is somewhat more liberal (Figure 3.3).

Looking at the IMT-Advanced documentation, the initial version of LTE is not capable of meeting the expected 4G performance values. As an example, LTE prior to the Release 10 is not able to provide the 1 Gb/s data rate in downlink, which is one of the requirements of IMT-Advanced. Yet, it is common to see the early version of LTE and sometimes even UMTS HSPA+ networks being called 4G.

The mobile telecommunications industry has typically interpreted basic LTE as 4G. The practical explanation of these statement could be that LTE, in fact, is much closer to 4G than to 3G, and thus, the term 4G has already been adopted in the marketing of LTE. Another reason is the relatively long time the standardization takes, and the industry has obviously wanted to distinguish between the older 3G era and the introduction of LTE. Third, markets have simply decided that the final word for the practical definition of term 4G is not written on stone, so it is only fair to have some level of freedom for interpretation.

Regardless of the classification of LTE as a last step of 3G, or as a first or pre-step in the 4G era, it paves the way for the actual ITU R–defined 4G, and latest by the deployment of LTE-Advanced providing the gigabit experiences, the networks can be claimed to represent fully compatible 4G system. Depending on the final role and success of WiMAX, this might actually be the first time in mobile communications history for the real merging of the solutions into a single solution, based on LTE.

It is good to remember that the 4G experience is a sum of many items, such as evolved antenna technologies, carrier aggregation, and new mobile device categories, so the highest data rates will be available for us consumers in a gradual manner. It is also important to note that the latest 4G performance requires the support of these functionalities from both network and user equipment. Meanwhile, it is, after all, not so big deal what performance is behind the term 4G, while the user experience proves clear differences between the older solutions and the new ones.

3.2.5 5G

3.2.5.1 Definition

Interestingly, as the utilization of term 4G has been used actively in the commercial pre-LTE-A Release 10 era, there seems to be inflation noted in the terminology, as some earlier systems were already called 5G. Nevertheless, the fully ITU-compliant 5G is still under development; the complete 5G deployment could take place around 2020. We thus need to wait to be able to experience the forthcoming performance of the real 5G.

Faster data rates, low latency, maturation of the IoT, and the ability to handle a massive amount of data are among the major benefits promised by 5G.

The term 5G is thus confusing. We keep hearing public announcements about the 5G network deployments being right around the corner while 4G deployment is still in its hottest phase – considerable amount of networks seemingly switched on just a moment ago, and hundreds of pre-4G LTE Release 8 and 9 networks were commercialized relatively recently.

Terminology that's been applied to mobile-communications networks through the third generation has been quite understandable, as 3G refers to a set of systems that comply with the IMT-2000 (IMT for 3G) requirements designed by the ITU. Thus, the cdma2000, UMTS/HSPA, and their respective evolved systems belong to the third generation as the main representatives of this era.

The definition of the fourth generation is equally straightforward, based on the ITU's IMT-Advanced requirements. Up to now, two systems have been compliant with this official 4G category: 3GPP LTE-Advanced as of Release 10, and IEEE 802.16m (also known as WiMAX2). The first Release 8 and 9 LTE networks were deployed in 2010–2011, and their most active commercialization phase took place around 2012–2014. Referring to ITU terminology, these networks prior to Release 10 still represent the evolved 3G era that, as soon as upgraded, leads the way to the "full-blooded" 4G.

While 4G is being deployed or upgraded from the previous LTE releases, 5G has already generated big interest. The year 2016 was "showtime" for many companies to demonstrate how far the technical limits can be pushed. Some examples of the already announced initiations include Verizon Wireless, which partners with technology leaders in two Verizon innovation centers, and Qualcomm, which has demonstrated the capabilities of LTE-Advanced Pro via millimeter-wave setup.

These examples, as well as many other demos and field trials indicated the performance and capacity increase that 5G may provide, although the commercial and commonly agreed form of the 5G has yet to be seen. Whenever ready for consumers, it can be assumed that the 5G era will represent something much more than merely a set of high-performance mobile networks. It will, in fact, pave the way for enabling seamlessly connected society with important capabilities to connect a large amount of ways-on IoT devices.

The idea of 5G is to rely on both old and new technologies on licensed and unlicensed radio frequency (RF) bands that extend up to several gigahertz bands to bring together people, things, data, apps, transport systems, and complete cities – in other words, everything that can be connected. Thus, 5G functions as a platform for ensuring smooth development of the IoT, and it acts as an enabler for smart networked communications.

The aim of the global 5G standard is to provide interoperability between networks and devices, offer high-capacity energy-efficient and secure systems, and significantly increase user data rates with much less delay in the response time. Nevertheless, prior to the 3GPP Release 16, the fifth generation represents a set of ideas for highly evolved systems beyond 4G, so its development still takes some time. As has been the case with the previous generations, the ITU has taken an active role in coordinating the global development of 5G.

3.2.5.2   5G Standardization

The ITU-R (radio section of the ITU) is the highest-level authority for defining the universal principles of 5G. The ITU is thus planning to produce a set of requirements for the official 5G-capable systems under the term IMT-2020. As the term indicates, the commercial systems are assumed to be ready for deployment as of 2020. This follows the logical path for ITU-defined 3G and 4G, as can be seen in Figure 3.2.

The IMT-2020 is, in practice, a program to describe 5G as a next evolution step after the IMT-2000 and IMT-Advanced, and it sets the stage for international 5G research activities. The aim of the ITU-R is to finalize the vision of 5G mobile broadband society, which, in turn, is an instrumental base for the ITU's frequency allocation discussions at the WRC (World Radio Conference) events. Up to now, WRC-15 was the most concrete session for discussing 5G frequency strategies. The WRC decides the ways in which frequency bands are reorganized for current and forthcoming networks, including those that will be assigned to 5G. The next WRC takes place in 2019, and that will be the groundbreaking milestone for the decision of the IMT-2020-compatible 5G systems' radio allocation strategies.

Concretely, the Working Party 5D (WP5D) of ITU-R coordinates information sharing about the advances of 5G. These include the vision and technical trends, requirements, RF sharing and compatibility, support for applications and deployments, and most importantly, the creation of IMT-2020 requirement specifications.

While the ITU-R works on the high-level definition of the 5G framework, one active standardization body driving for practical 5G solutions is the 3GPP, which is committed to submitting a candidate technology to the IMT-2020 process. The 3GPP aims to send

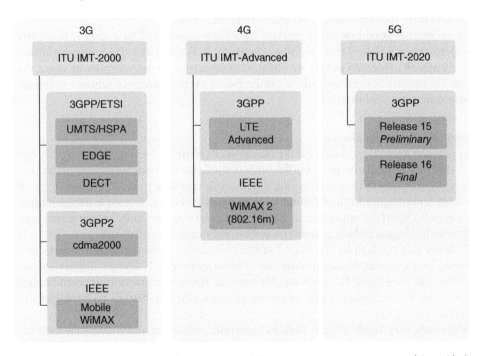

Figure 3.2 The main systems per 3G, 4G, and 5G mobile communications systems complying with the respective ITU requirements.

the initial technical proposal to the ITU-R WP5D meeting #32 in June 2019, and plans to provide the detailed specification by meeting #36 in October 2020. To align the technical specification work accordingly, the 3GPP has decided to submit the final 5G candidate proposal based on the further evolved LTE-Advanced specifications as of December 2019. In addition to the 3GPP, there may also be other candidate technologies.

As for the 3GPP specifications, 5G will have an impact on several technology areas that most affect the radio interface. The expected aim is to increase the theoretical 4G data rates, perhaps 10–50 times higher, while the data response time would drastically reduce to near zero.

The 3GPP RAN TSG (Radio Access Network Technical Specification Group) is the entity committed to more specifically identify these requirements, scope, and 3GPP requirements for the new radio interface. The RAN TSG works in parallel with the ongoing LTE evolution that belongs to the LTE-Advanced phase of the 3GPP, aiming to comply with the future IMT-2020 requirements of the ITU. At the same time, the evolved core network technologies need to be revised by the system architecture teams so that they can support the increased data rates accordingly.

3.2.5.3 Research
At present, many ideas are circulating about the form of the final 5G. Major operators and device manufacturers are conducting technology investigations, demos, and trials aimed at proving the concepts and contributing to the standardization. These activities are beneficial for the overall development of 5G, but they represent proprietary solutions until the international standardization ensures the jointly agreed upon 5G definitions, which, in turn, facilitates global 5G interoperability [4–6].

In addition, several research programs have been established to study the feasibility and performance of new ideas at the academic level. For instance, the European Union coordinates 5G research programs under various teams. More information about the latest EU-funded 5G research plans can be found on the European Union's web page that summarizes 5G initiatives.

3.2.5.4 Challenges for Electronics
One of the expected key abilities of the 5G networks is the high energy efficiency to cope with the huge array of low-power IoT devices in the field. The benefits include better cost efficiency, sustainability, and widening the network coverage to remote areas. Some of the base technologies for facilitating the low energy include advanced beamforming as well as radio interface optimization via user-data and system-control plane separation. Other technologies include reliance on virtualized networks and clouds.

Systems also need to be developed at the component level for both networks and devices. In particular, autonomously functioning remote IoT devices require special attention, as they must function reliably without human interaction or maintenance perhaps for several years. Thus, advances in more-efficient battery technologies are essential.

Moreover, very small devices such as consumer wearables and M2M (machine-to-machine) sensor equipment may require much smaller electronic component form factors, including tiny wafer-level subscriber modules that still comply with the demanding reliability and durability requirements in harsh conditions. At the same time, the need for enhanced security aspects will require innovative solutions at the hardware and software levels.

3.2.5.5   Expected 5G in Practice

The industry seems to agree that 5G will be a combination of novelty, yet-to-be-developed, and standardized solutions together with existing systems, including commercially available mobile generations that jointly result in increased data rates and capacity. Some of the 5G use cases could involve the support of tactile internet and augmented virtual reality, which provide completely new, fluent, and highly attractive user experiences.

Resulting from a long period of mobile-communications development, 5G's roots can be traced back to the 1980s, when first-generation mobile-communications networks began to be reality. Ever since then, the newer generations up to 4G have been based on the earlier experiences and learnings, providing a base for developers to design enhanced technologies for the access, transport, signaling, and overall performance of the systems. However, the telecom industry has identified a great need for considerably faster end-user data rates to cope with the demands of the evolving multimedia. To that end, 5G would be able to handle these challenging capacity requirements to provide fluent user experiences even for the most advanced virtual-reality applications. At the same time, the exponentially growing number of the IoT devices require new security measures (see next section), including potential security-breach monitoring and prevention.

Along with the new M2M and IoT applications and services, there will be role-changing technologies developed to support and complement existing ones. In that regard, 5G is one of the most logical bases for managing this environment, in combination with legacy systems.

Although 5G is still a set of fragmented ideas (until the ITU officially dictates its requirements and selects suitable technologies from among the candidates), 5G systems will soon become a reality. Thus, we can expect to see many novelty solutions, ranging from highly integrated wearable devices, household appliances, industry solutions, robotics, and self-driving cars to virtual reality and other advanced, always-on technologies that benefit greatly from enabling 5G platforms.

In addition to the "traditional" types of IoT devices, such as wearable devices with integrated mobile communications systems, car communications systems, and utility meters, there are also emerging technology areas, e.g. self-driving cars, that require high reliability for the functionality as well as for the secure communications – both requirements that can be met with 5G.

3.2.5.6   5G and Security

In terms of security assurance in the new 5G era, impacts can be expected in the "traditional" forms of SIM/Universal SIM (UICC, universal integrated circuit card) and subscription types in the much more dynamic environment. Ongoing efforts in developing interoperable subscription-management solutions that respond in near real time when users opt to change devices and/or operators are creating one of the building blocks for the always-connected society.

It's still not clear how the consumer and M2M devices will look physically in the 5G era, but we can assume that there will be much more variety compared to any previous mobile-network generation. This includes multiple wearable devices per user and highly advanced control and monitoring equipment.

These completely new types of devices can change the role of removable subscription identity modules such as SIM/UICC; the much smaller personal devices require smaller form factors. At the same time, the techniques to tackle with the constantly changing subscriptions between devices need to be developed further as do their security solutions. Cloud-based security, such as tokenization and host card emulation (HCE), as well as device-based technologies like trusted execution environment (TEE), may dominate the 5G era. However, the traditional SIM/UICC can still act as a base for the high security demands.

3.3 The Role of 3GPP in LPWA and IoT

This section outlines examples of the IoT deployments and estimates current penetration and outlook for near future. First, an overview of updated standards is presented. The standards are defining the global cellular-based narrow-band IoT solutions while the more limited IoT connectivity solutions are provided by proprietary providers. Second, deployment examples are presented for LPWA (low-power wide area) IoT connectivity solutions.

LPWA has two subcategories: (i) current proprietary LPWA technologies, such as Ingenu's random phase multiple access (RPMA), Sigfox, and LoRa, operating typically on unlicensed spectrum; (ii) 3GPP-standardized cellular IoT (C-IoT) technologies, operating typically on licensed spectrum. The early C-IoT is provided, e.g. via GSM (Release 8) on 200 kHz band and up to 0.5 Mb/s UL/DL (uplink/downlink), or EC-GSM-IoT as of Release 13 offering 200 kHz band with up to 140 kb/s UL/DL.

The LTE standards further optimize the C-IoT via Cat. 1, M1, and NB1 on the path toward the massive IoT (mIoT) as defined in 5G Release 16.

3.3.1 Overall LPWA Market Landscape

The LPWA market is growing rapidly. Compared to 2015, there is a considerable growth in the major LPWA solutions, which as (i) unlicensed spectrum: Ingenu, LoRa, Sigfox, Weightless, and (ii) licensed spectrum: cellular LPWA, i.e. NB-IoT, EC-GSM, LTE-M (see Table 3.2).

Two-thirds of the LPWA deployments in 2015 took place in Western Europe and the United States. Nevertheless, these regions represented a third of the new initiatives in 2016. Asia–Pacific region is the main driver of the posterior regional shift, representing almost one-third of all new network initiatives in 2016. Mobile network operator, MNO-driven initiatives are strong in Japan, Singapore, and South Korea, and many initiatives rely on NB-IoT technology (e.g. M1 Singapore and KT South Korea) while, e.g. SoftBank Japan has LoRa implementation.

Start-ups having LoRa and Sigfox are still strong players. Australia and New Zealand launched nine different LPWA initiatives.

North America has been the largest LPWA region thanks to the adoption of RPMA devices in predominantly private networks. LoRa and Sigfox networks have also been actively rolled out in the United States. Western Europe became the largest region in

Table 3.2 Comparison of LPWA systems.

LPWA variant	Status
LoRa	• LoRa implementation in NA has presented with challenges imposed by FCC regulations. [7] Hybrid mode would solve this challenge. • June 2016: 100 US cities covered by Senet LoRa network for IoT. [8] • April 2016: LORIOT expands the LoRa network in California. [9]
Sigfox	• Another major player in the global and domestic IoT network space is French company Sigfox, which is in the process of a 10-city US deployment: San Fransisco, San Jose and Los Angeles, California; New York; Boston, Massachusetts; Atlanta, Georgia; Austin, Houston, and Dallas, Texas; and Chicago, Illinois [8]
RPMA	• February 15, 2017: Ingenu teaming with Microsoft to utilize the Microsoft Azure IoT Hub to facilitate the deployment of end-to-end IoT based on RPMA. Azure IoT Hub supports various OSs and protocols, providing IoT solution developers with the means to connect, monitor, and manage IoT assets [10]
LTE-based	• T-Mobile and Vodafone active in NB-IoT deployments in Europe. [11] • Huawei has announced that it will commercialize its NB-IoT framework and technologies early in 2017, including the release of its System on a Chip (SoC) commercial processor, Boudica. Nevertheless, the company is still little known in North America [12]

terms of LPWA connections in 2017, but Asia Pacific looked to overtake it in 2018, and by 2020 the latter is expected to represent nearly 46% of the total.

The US government is investing significant amounts toward the implementation of IoT across various sectors such as infrastructure and utilities under programs such as Smart America. The US government offers a supportive environment for research and development, which is facilitating advancements in IoT applications.

An increasing number of unlicensed LPWA and cellular (LTE-M) networks coexist. AT&T and Verizon have committed to nationwide LTE-M networks and a number of companies are building LPWA networks in unlicensed bands, most notably Comcast using LoRa. In Europe, there are also plans for LTE-M deployments to complement LoRa networks (Netherlands). This could be increasingly popular mode in the near future, i.e. the combination of unlicensed licensed networks.

MNOs are under pressure to grow IoT revenues, and there might be acquisitions seen. Up today, IoT revenue of MNOs has not typically exceed 1%. In 2025, MNO IoT revenue is expected to exceed 5% (20% annual growth rate estimated). As an example of the transition and preparation for the new era, Verizon had spent over US $3 billion in 2016 by acquiring Fleetmatics, Telogis etc. Narrow Band cellular-based IoT technologies will thus likely create a new global market opportunity in the space for LTE operators [13].

3.3.2 General IoT Requirements

The IoT is remarkably meant for LPWA [14]. The requirements for such model are:

• Wide coverage, including rural areas and indoors.
• Terminals that are optimized as power efficiency. The consensus of industry indicates minimum battery duration of 5–10 years.

- Low costs. It may be estimated that a typical module for connectivity should be less than US $5.
- Good resistance for transmission delays, and minimum latency.
- Enhanced downlink transmission capability. The IoT should support downlink configuration and remote upgrade of terminal SW in order to optimize operation and maintenance costs.

Whilst there are several proprietary LPWA technologies, also 3GPP defines intermediate step toward 5G networks as the current 4G is evolving to better support narrow-band IoT connectivity.

The 3GPP standardization provides IoT connectivity solutions via LTE which are LTE Cat 1, LTE Cat-M1 (eMTC, evolved machine-type communication), and LTE Cat NB1 (Narrowband IoT). These modes aim to tackle the varying needs of IoT environments for existing licensed and unlicensed low-power wide-area networks, and the goal is to provide reliable and energy-efficient data transmission offering billions of IoT devices. The LTE-based standards are competitors with Sigfox, LoRa and Ingenu.

3.3.3 LTE-IoT Mode Comparison

Figure 3.3 and Table 3.3 summarize the IoT category development in LTE.

3.3.4 Statistics and Projections

It can be forecasted that the most popular LPWA applications are related to smart meters (about 45% of LPWA connections in 2020). Industrial and financial applications

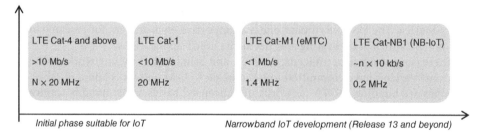

Figure 3.3 The main parameter comparison of 3GPP NB-IoT modes.

Table 3.3 Comparison of some 3GPP IoT modes.

	Cat 1 LTE Release 8	Cat 0 LTE Release 12	Cat M1 LTE Release 13	NB-IoT LTE Release 13
DL peak data rate (Mb/s)	10	1	1	0.2
UL peak data rate (Mb/s)	5	1	1	0.2
Duplex	Full	Half/Full	Half/Full	Half
UE bandwidth (MHz)	20	20	1.4	0.2
Max. transmit power (dBm)	23	23	20/23	23
Modem complexity	High	Medium	Low/Medium	Very low

will also be popular, followed by consumer electronics with high growth rates in 2020. Smart city applications such as intelligent lighting and parking, as well as smart buildings, are increasingly popular forming estimated 12% of LPWA connections in 2020. Nevertheless, the adoption of LPWA in land vehicle-based applications is expected to be limited.

3.3.5 Standardization

3.3.5.1 3GPP

3GPP completed the standardization of NB-IoT on June 2016. The new narrowband radio technology is being developed for the Internet of Things (IoT), implementing the feature into Release 13 (LTE Advanced Pro) specifications. The portfolio of 3GPP technologies for MNOs provide solutions for diverse markets and includes NB-IoT, eMTC, and EC-GSM-IoT. The complete list can be found in Ref. [15].

The eMTC is delivering further LTE enhancements for Machine Type Communications, building on the work that started in Release 12 (UE Cat 0, new power saving mode). NB-IoT is a new radio added to the LTE, optimized for the low end of the market. EC-GSM-IoT brings enhanced GPRS (EGPRS) additions, making in combination with Power Save Mode (PSM) also GSM/EDGE systems IoT-ready.

3.3.5.2 LoRa

LoRa Alliance creates standards for the IoT and promotes to the growing ecosystem for driving volume deployments for low power wide area networks (LPWANs). According to LoRa Alliance, the LPWAN's are projected to connect 50% of the predicted IoT volumes.

LoRa Alliance is standardizing LPWAN with the LoRaWAN specification and has interoperability certification and compliance program. LoRaWAN devices will be able to be deployed in multiple networks and roam from one network to another irrespective of network infrastructure or operator.

LoRaWAN is a specification, which is standardized by the LoRa Alliance that includes ST, Microchip and Amiho. It is based on technology developed by chipset supplier Semtech [16]. It is being deployed e.g. cross Europe.

In practice, in Europe, the LoRa gateway needs merely one SX1301 concentrator, which supports up to eight separate bands. The European Telecommunications Standards Institute (ETSI) requires LoRa to support minimum three channels at 868.10 MHz, 868.30 MHz, and 868.50 MHz.

In North America, though, Federal Communications Commission (FCC) defines 64×125 kHz channels and 8×500 kHz channels added by 8 downlink channels; nevertheless, FCC regulation requires hopping over at least 50 channels when maximum output power is applied. It would be possible to have user equipment with less channels (6×125 kHz channels) if the device transmit power is limited to 21 dBm. This, in turn, complicates the HW (compared to Europe) [17]. For this reason, LoRa deployments in NA region are limited.

3.3.5.3 Ingenu

Ingenu is building the Machine Network, i.e. IoT network dedicated to LPWA connectivity for machines [18].

The RPMA of Ingenu is designed for low-power, low-cost, long-range applications where battery life and network longevity are of importance. The technology is meant for benefiting especially utility devices in wide areas in such a way that the transmission is protected and guaranteed. The RPMA is based on proprietary solutions of Ingenu.

3.3.5.4 Sigfox

The aim of Sigfox is to provide IoT connectivity solution via a dedicated network. The marketing attributes of Sigfox include inexpensive, reliable, and low-power solution for connecting sensors and devices. The overall requirements for the technology include simplicity, referring to no need for configuration, connection request or signaling, or autonomy, which means extremely low energy consumption allowing years of autonomy on a single battery charge, and small payload, which results in optimal assets or multimedia via the need for only small messages.

3.3.6 Deployments

3GPP has been active in the standardization of the C-IoT after the first phase of the non-IMSI-based LPWA technologies. NB-IoT is expected to be popular, providing reliable connectivity, it consumes low power, and has cost benefits being possible to be integrated into the existing infrastructure on MNOs. Most IoT devices, especially sensors, do not send data of large sizes, which justifies the narrow band for data connectivity over a long range. The deployment of NB-IoT in sub-1 GHz bands enables adequate propagation and penetration characteristics of NB-IoT.

According to publicly available statistics, there are many countries considering or deploying LoRa, especially in sensor environments, including smart agriculture (moisture sensors), intelligent harbor operation system, and LoRa provides base for various local and global initiatives for investigating the suitability of it in different business models, such as offering of gateway on a crowdfunding platform such as Kickstarter or Indiegogo [19].

RPMA is being deployed especially in the United States. There are many initiatives for investigating RPMA, e.g. in sensor networking of agriculture environment; Smart City environmental monitoring, etc. [12]. RPMA is in approx. 30 cities in the United States by the end of 2016.

Sigfox aims to deploy the solution in the United States and globally. As an example, based on [20], Sigfox achieves record growth in the United States, confirms network coverage in 100+ US cities, and in many other countries. Sigfox is cooperating with various chipset vendors such as STM and NXP.

3.3.7 Comparison of LPWAs

In the initial phase of noncellular LPWA technologies, their main arguments have been that LPWA complements existing cellular mobile network and short-range technologies. They also are said to enable wide area communications at lower cost points, and their power consumption has been more optimal that those of cellular based. Also, the low costs of wide area out-of-the box connectivity would allow more business-cases and greater freedom in terms of deployment locations.

Along with the further development of the LTE-based IoT, the relevance of these arguments is lowering, making both cellular and noncellular IoT solutions more comparable. According to [14], the benefit of all the 3GPP-based cellular IoT variants (low-cost LTE with machine-to-machine -type of communications [MTC] in Release 12; eMTC in Release 13; and NB-IoT in Release 13) are superior as for the security, ecosystem, network cost (TCO, total cost of ownership) and spectrum compared to the other LPWA technologies. This is due to the developed procedures of the LTE network for the authentication, authorization as well as UE's HW-SE-based encryption.

In the initial phase of LTE-based IoT devices, the module cost does not compete with the other LPWA devices but as NB-IoT takes place, the cost is adjusting to more comparable levels. Also, the service coverage of the LTE-based IoT will increase gradually, MTC providing lowest coverage whilst the coverage of the eMTC is comparable with other LPWAs, and NB-IoT exceeding the coverage.

3.3.8 Security Aspects of IoT Devices

If no HW-SE involved, the main question is how to provide credentials to devices to identify them securely and in a unique way. It is also important to reliably identify sensors and devices and to ensure from scratch that the identity cannot be cloned.

C-IoT is based on well-known security principles of ETSI and 3GPP; currently widely expanding non-C-IoT ecosystems bring along with a variety of new companies into the markets. The security expertise and experience related to the noncellular IoT are not necessarily up-to-date while IoT opens exceptionally broad attack vectors, seriously jeopardizing the security of the ecosystem.

Relevant knowledge needs to be ensured amongst users and network providers when combining noncellular LPWA elements, data, and applications. Some degree of isolation or disabling system, device, and application functionalities may be required to provide adequate protection.

Security procedures as per 3GPP technical specifications provide the following benefits, such as the fact that (e)UICC communicates via secure channels with the core network. There also is a native mutual authentication integrated into the network, and the cellular-based LPWANs are interoperable, based on internationally approved 3GPP standards.

According to [21], both LoRa and Sigfox offer some comparable security functions. Sigfox is using encryption mechanism for authenticating the device but is not encrypting communications that need additional solutions on firmware. Sigfox is based on a network authentication plus a frequency hopping algorithm and sequence number, avoiding a third-party device to emit with original device identity. LoRa is using three different keys to encrypt at device, application, and network level. This ensures that when you have multiple network, public and private, operating the same frequencies, they do not process data from other one (i.e. so a third party cannot listen on your device data).

Source [15] further describes that the protection of the communications refers to potential attacker disturbing or blocking the messages by avoiding reception via interfering noise at selected elements. Sigfox and LoRa are not fully equivalent on this point as LoRaWAN requires the object to receive the network channel configuration in the connect sequence. Once LoRaWAN receives the initial information, it can communicate on its own. The connection sequence is thus initiated on the first communication, which is the vulnerable moment. Source [21] concludes that e.g. valuable goods tracking is not a suitable application for LoRaWan due to this vulnerability.

The Machine Network of Ingenu's RPMA solution is marketed capable to assure high reliability and security. The high level of security refers to the protection of the privacy of the data as well as the ensuring of the delivery. The security of Ingenu is built in, with 256-bit encryption, 16-byte hash, and two-way authentication. The solution utilizes heavy coding rate (1/2), which is suitable for ensuring the reliability of the message delivery over low bit rates.

Both proprietary (or, internationally limited) specifications of LPWA variants such as LoRa, RPMA, and Sigfox, as well as globally standardized LTE-based IoT solutions, are advancing, and respective deployments are ongoing in international level. Even if they are mutually incompatible, each solution seems to have its market share both in the United States as well as in the global environment.

As for IoT devices, the widest interoperability is provided by the variants of the 3GPP LTE IoT solutions, while other technologies are somewhat more limited. 3GPP has provided intermediate solutions, and the further specification work toward 5G further optimized the LTE-based IoT connectivity, making it very attractive option for IoT clients.

Nevertheless, it can be estimated that typical IoT M2M devices, once deployed, do not need to be changed often so the utilization of any of the available technologies is justified. As the C-IoT is evolving, the performance and cost of modules will be comparable with those of other LPWA variants. The better security level of C-IoT will thus be one of the most important differentiators.

Today, some of the most concrete competing solutions are currently under deployment: Sigfox, Ingenu's RPMA, nWave, LoRa, and possibly Dash7, while the 3GPP-based C-IoT is becoming a more feasible option as for device cost, coverage, and other attributes that are already becoming comparable with or exceeding other LPWAs. As for security, the C-IoT provides the strongest protection while the somewhat-less-protected level of the noncellular LPWA may be justified for many of their environments as such or by adding an extra layer of security.

Nevertheless, Ref. [22] states that the current non-LPWA solutions are fragmented and nonstandardized, the negatives including poor reliability and low security, with high operational and maintenance costs. The new overlay network deployment is complex. Ref. [22], among similar ones, is in favor of C-IoT, in particular NB-IoT.

Some key security standards relevant to IoT in North America are the following:

- NERC CIP 002-009 cyber security framework for critical cyber assets
- NIST SP 800-53 guidelines for protecting critical cyber assets
- FIPS 140-2 Level 2 encryption standards
- NISTIR-7628 guidelines for smart grid cybersecurity

3.3.9 What Will 5G Offer in IoT Landscape vs. Non-IMSI LPWA?

The IoT is developing in a parallel fashion with the first phase of 5G. Thus, all the modes developed so far live their commercial phase as they are being adapted into the LTE service set of operators. The market evolves, but it can be argued that the Cat-M and NB-IoT modes may be the most widely deployed soon. The Release 15 5G networks rely on these IoT modes of 3GPP.

As the Release 16-based networks are deployed, there can be expected to be a native mIoT mode to support increasing amount of simultaneously communicating IoT devices. According to the ITU's IMT-2020 requirement set, the mIoT pillar can be expected to support up to 1 million devices per square kilometer areas.

3.4 The Role of 5G in Automotive (V2X)

Vehicular communication system is a network that consists of vehicles and roadside units. These represent communicating nodes and they deliver information between each other. Typical data include safety warnings and traffic information.

Vehicle-to-vehicle (V2V) is vehicular communication environment, which refers to the direct links between vehicles. Vehicle-to-infrastructure (V2I) involves other components beyond those located at vehicles, e.g. roadside elements. Vehicle-to-pedestrian (V2P) refers to the communications between vehicles and nearby persons outside of the vehicles. Vehicle-to-everything (V2X) consists of components of V2V, V2I, and V2P.

V2V communications is a special data transmission system referred to as two-way dedicated short-range communications (DSRC) with range of up to 300 m between devices. DSRC works at 5.9 GHz, the bandwidth being 75 MHz. This band has been allocated by, e.g. FCC for use by intelligent transportations systems (ITS) vehicle safety, and mobility applications. Wireless access in vehicular environments (WAVE) are a set of standards within the DSRC suite to allow cooperative and safety critical applications to be supported.

Also, cellular components are being developed, with LTE evolution offering a functional base. V2X and its variants form a part belonging to the ITS concept. The vehicular communication systems are designed to avoid accidents and traffic congestion.

V2V, also referred to as VANET (vehicular ad-hoc network) or car-to-car communications, supported by V2I communication, has been designed for providing automatic communications between vehicles. V2V is being developed by various automotive firms such as General Motors, Audi, BMW, Daimler, Honda, and Volvo. In addition, there are various associate members such as LG, Siemens, and Atmel. Car-to-Car Communication Consortium (C2C CC) is developing the overall system.

The aim of the C2C CC is to enhance road safety and traffic efficiency via Cooperative Intelligent Transport Systems and Services (C-ITS). There are many applications for V2V, e.g. navigation and safety apps. The United States has been planning to make V2V mandatory, possibly by 2020. The communications security of V2V relies currently on public key infrastructure (PKI).

3.5 The Role of 5G in the Cyber-World

3.5.1 Introduction

We are firmly moving toward always-connected society that provides access to practically unlimited sources of information at any time regardless of our location. This is a result of the popularity of wireless communications, which has converted to a daily necessity for major part of the world's population. As a concrete example of this environment, the personal messaging is increasing exponentially for sharing information, and has raised a totally new user generation that is less concerned about personal privacy – the new technology has, in fact, sculptured our societies and cultures all over the globe by making it extremely easy to communicate with anybody regardless of the social status.

Often, even the most personal instances are shared with the rest of the world. The wide deployment of new techniques and new ways of using them have opened a great window for such an amazing way of communications that provide us now easy ways to spread news all over the world practically in real-time. At the same time, the increasing openness and technological advances may also expose important security holes that jeopardize confidential information – even the most extrovert users may not want to expose, say, all the identity and bank account details for everybody.

Not only the personal communications culture is facing drastic changes, but the overall information society is under major facelift, too. The IoT is developing with huge leaps, which will provide means for always connected machines and services to communicate autonomously between other entities and data networks to ease our own daily life.

3.5.2 Cyber and IoT

The IoT is as versatile as the word *things* sounds. It refers to a vast set of all kind of equipment connected to Internet including surveillance cameras, refrigerators, printers, sensors, self-driving cars, and telematics devices that are able to communicate and perform actions. The key idea of the IoT devices is thus to facilitate useful functions via the always-on connectivity, which provides fluent user experiences and makes our lives more enjoyable and efficient in the transition from the information society toward the truly connected society.

The IoT provides vast amount of new possibilities for managing, coordinating, automatizing and benefiting from M2M communications without human intervention, but it also relies on human interactions when needed. The IoT is, in fact, an object that can represent itself digitally and becomes something greater than the object by itself. In other words, the object will be connected to surrounding objects and databases so that it forms part of a whole environment. This is fascinating because as soon as multitudes of such objects are acting in unison, they start having *ambient intelligence*. One example of such an environment is a sensor network that may detect a car accident that starts blocking a highway. The sensors may pre-digest information and send it to relevant cooperating organizations such as airlines, train companies, and hotels, which together can reason that there may be important changes coming soon due to the accident, so all of these stakeholders can prepare themselves by increasing capacity for alternative transportation and lodging in the area. This is also one of the use cases related to Big Data and data mining that, when properly analyzed and assessed, can provide huge benefits for everybody, as it is possible to adjust the environment practically in real time based on the varying incidents.

As a result of the active development in these technology areas, totally new type of devices will be introduced to the consumer market as a base for even further, highly innovative services. In addition to such a huge amount of new opportunities, this also opens known and totally new security threats that may compromise the identity of the users and confidentiality of the information, which, in turn, may jeopardize the safety of, e.g. economic funds. The security breaches may even threaten personal well-being of humans, as in the case of compromised medical or traffic control applications. One of the concrete demonstrations of such life-threatening risks is related to the possibility for remotely taking over control of a connected, self-driving vehicle.

The protection of citizens against such threats is more than challenging in the modern environment with increasing level of sophisticated cyberattacks. This can be addressed by increasing cyber surveillance, the entities being individuals, governments, and whole ecosystem of banks, production chains, armed forces, and educational institutes. The IoT can thus also be interpreted as an elemental component in the cyber-world, and the proper level of knowledge of the cyberattacks and respective protection methods is needed by everybody for surviving in this new environment.

3.5.3   What Is Cyber?

Cyber refers to a very wide environment. It may consist of many familiar components from our daily life such as Internet, mobile and fixed telephony, and other topics that are related to computing and connectivity to internal and external data networks. In addition to the environment as we are capable of observing it, *cyber* is also something to do with hidden secrets, their protection, and intentions to attacks for stealing or exposing them. Furthermore, cyber can be related to the artificial intelligent – capable and autonomously functioning machines such as moving robots. *Cyber* is, in fact, already present at all the levels of our society including personal life, governments, businesses, banks, and defense forces. Remarkably, the rapidly growing environment of IoT belongs as an integral part into cyber.

> "What are you?"
> "I already answered that," snapped the machine, clearly annoyed.
> "I mean, are you man or robot?" explained Klapaucius.
> "And what, according to you, is the difference?" said the machine.
> (Stanislaw Lem, *The Cyberiad* stories)

The current IoT can be understood as a set of both passive and active physical objects that are part of the information network in a fluent and seamless way. These objects participate in the functioning of whole systems and contribute to the actions of various areas of everyday life, such as learning, health care, telematics, and businesses. The IoT is paving the way to a completely novel society that integrates the "traditional" physical world as we see and feel it, the digital world with the information stored and transferred in bits and bytes, and the virtual cyber-world.

Just as for the term *cyber*, the term *cybersecurity* is broad and may include basically all kind of security breaches such as cyberattacks and respective protection mechanisms. There are thus confusingly many definitions for the topic. One of the intentions to summarize it is presented in the *Tallinn Manual on the International Law Applicable to Cyber Warfare*, [23] which describes the meaning for the cyber operation from the military point of view. According to this source, one of the very key aspects of cyber warfare is that it may cause injury or death to persons, or damage or destruction to objects.

Furthermore, a broad spectrum of cyberattack types could include intrusion, surveillance, recording of data, espionage, extraction, destruction and manipulation of data, theft of intellectual property, control of devices and systems, kinetic effect through control of devices, destruction of devices, property and critical infrastructure, individual lethal effect, and operations with national impact. Cybersecurity, in turn, provides the countermeasures for the cyberattacks.

In general, the cyberattack can be categorized into three components, which are intelligence (for gaining access, executing a cyber payload, and understanding the target environment), weapons (typically target-specific and activated under specially planned conditions), and calculated human decision. The combination of these elements forms a cyber force.

As for the human point of view, there are two types of consequences from the cyberattack: the ones that are nonviolent yet impactful for the functions of information society (so-called virtual destruction and disruption) and the ones that lead in physical injuries and destruction. As a classical example of cyberattack from 2010, the Stuxnet virus infected control units of targeted nuclear centrifuges, causing them to accelerate and self-destruct. This case is a concrete proof of the magnitude of the consequences the potential attacks may lead to. In addition to the physical damage, they also can result a total and permanent destruction of information storage.

Cyberattacks are reality, and their role will be much more important in the forthcoming years. Not only will they have impact on the new role of defense forces, but also it will be highly relevant in daily life. The unprotected environment may lead to disastrous consequences for individuals and organizations, and may even affect the economical foundations of the societies. Thus, all levels from the individuals' home environment with the simplest IoT devices up to the most professional cyber army need to deploy, use, and maintain cyber-protection mechanisms.

3.5.4 Standardization

Cyberthreats are so different and advanced compared to the "traditional" hacking efforts that many standardization and regulation bodies currently take into account in the development work.

As stated by the ISO (International Standardization Organization), cyberthreats continue to plague governments and businesses around the world. These threats are on the rise as cybercriminals increase their focus and know-how. Thus, ISO is one of the entities charged with tackling the issue in international level. As an example, ISO/IEC 27001 provides a management framework for assessing and treating risks, including the cyber-oriented attacks that can damage business, governments, and even national infrastructures. The ISO also produces other standards that are related to cyber, such as ISO 15408, which is referred to as Common Criteria.

Another example of the major standardization bodies that take cyber into account is the ETSI, which established the Cyber Security Technical Committee (TC CYBER) in 2014. Its aim is to tackle with the growing threats of Internet communications by ensuring adequate security solutions, and it also provides a centralized point of expertise for other stakeholders in the standardization field.

Also many other standardization organizations have taken cyber as an important part in the further development of technical solutions in telecommunications, information technology, and other relevant realms. Some examples of these entities are NERC (North American Electric Reliability Corporation) especially for the critical infrastructure protection, and NIST (National Institute of Standards and Technology), which has produced a set of publications emphasizing the importance of security controls, principles, and management. Some other key contributors are the IETF (Internet Engineering Task Force), which, among other related documents, has produced a *Site Security Handbook* in form of the RFC 2196, and the ISA/IEC (International Society for Automation/International Electrotechnical Commission), which has produced the 62443-series of standards defining procedures for implementing industrial automation and control systems (IACS) in a secure manner.

3.5.5 Origin and Development

With such a diversity of definitions for cyber, one might wonder what the origin of the word is. In fact, if the situation is confusing now, with almost every type of words suffixed with cyber (e.g. cyberpunk, cyberspace, cyberwar …) – indeed in such a scale that the term has experienced a great inflation since the 1990s – it really was not clear back then, either.

The ancient explanation comes from Greek word "κυβερνᾶν," referring to the governing or steering. This merely political term by that time is quite far from the meaning cyber started to take finally around in the middle of twentieth century, in form of "cybernetics". The reason for adapting such a term was the intention to describe the investigations of those days about communication and control systems in living beings and machines. These studies involved highly cross-scientific research in engineering, biological, and social domains, and the term made a lot of sense to describe this cooperation.

Soon after, the term started to raise interest for describing all sort of technical advances in a futuristic way yet without clear definitions of the real meaning of those. This was the case especially in popular cultures such as science fiction movies and literature. Amongst many classical pieces, a great example of the products of this era is "The Cyberiad" stories by Stanislaw Lem, which contains fascinating novels involving robots and artificial intelligence (AI). Another variant was the Borg of *Star Trek* movies referring to pseudo-species of cybernetic beings, or cyborgs. Thus, cyber was also referring to as a combination of human behavior and robotics, in a kind of androids that were capable of communicating, reasoning, and making decisions.

In fact, the current development of machines equipped with AI has raised some serious concerns about the safety issues even in the highest scientific level. Professor Steven Hawking described concerns for the potential failure in controlling the machines that may eventually outsmart us.

> Success in creating AI would be the biggest event in human history. Unfortunately, it might also be the last. [24]

Although there are certainly issues to be worried about in the future of cybernetics, we are still in the relatively early stages of such development. Nevertheless, it would be important to agree to international rules for the further development of the AI to avoid such horror sceneries we have already seen in the most pessimistic sci-fi movies with machines attacking people. On the other side, there also are plenty of opportunities that can help humans in many areas, so it would not be any miracle to see soon the first robots roaming the streets to serve us by carrying merchandise from local groceries.

To return to the definitions, the cyber tried to take shape especially in 1980s and 1990s with such an excess of cyber-terms that it started to experience already some sort of inflation. Some of the intentions to establish cyber as a term were related even to the remote adult relationships that, after a short hype lost soon the grip as an idea. In general, *cyber* was used as a synonym for so many things that they could easily be replaced by terms such as "digital," "advanced" or "future." One example, as noted earlier, could be the term *cyberpunk*, which referred to merely more futuristic and electronic sound compared to the previous variants of the genre in question. After all, these somewhat unsuccessful intentions to adopt the term into such a variety of environments, cyber has had more traction in the military environment, as well as in describing threats and security issues of the Information and Communications Technologies (ICT).

Furthermore, it can be claimed that advanced robotics is one of the important building blocks of cyber. Some of the recent examples of this domain are the animal-shaped and human-kind robots developed by Boston Dynamics. It could be interpreted that such type of robots, equipped with long autonomous functioning and AI could be a subset of cyber, especially in the military environment, yet useful in many other domains, too.

As promised in the previous section, let's continue seeking for the answer to the actual definition of cyber. As many standardization bodies have taken cyber as a study topic even by using it in the respective team naming, the question is whether there is a standardized statement defining cyber at all. Now, this seems to be the tricky part, as many standards groups take the meaning of cyber for granted, typically assuming it as a very broad environment, as discussed previously. So perhaps it is safest to refer to some of the internationally recognized dictionaries and authorities of languages. Would it be actually too easy to rely on the Oxford dictionary, which states that cyber is of, relating to, or "characteristic of the culture of computers, information technology, and virtual reality"? [25] This certainly sounds logical, although it is frankly speaking so ample that it gives room for almost infinity of interpretations. It really does not help too much to know that the same source states it is originated from cybernetics, which is the science of communications and automatic control system in both machines and living things.

Let's try then via other language areas to understand if there are any more concrete statements to be found. The Spanish Real Academia Española states, loosely translated, that cyber refers to a prefix that is shortened version of cybernetics, and it forms part of terms related to world of computers and virtual reality.

So, the conclusion of our short study reveals that there is a huge amount of interpretations for describing cyber, the typical statements indicting it has something to do with computers, virtual reality, machines, and/or living things. After all the "trials and errors" in the efforts of getting cyber, and cyber-prefix into our vocabulary, there are still some a bit more systematically and constantly utilized terms such as cybersecurity and cybercrime.

3.5.6 Machine Learning and AI in 5G

The machine learning and artificial intelligence are estimated to form an increasingly important base for the 5G networks. As an example, the highly complicated and dynamic functions such as network slicing require automated actions that can contribute vastly to the optimal performance of the system. Another example is the self-optimizing network concept, which can benefit from fast responses of artificial intelligence in recovering from faulty situations and in adjusting the network performance based on the current need.

A variety of mobile communications stakeholder are considering and developing thus the technologies, which may be deployed to 5G as they mature.

References

1 J. Penttinen, "What's the Story with 5G?," 29 June 2016. http://electronicdesign.com/print/communications/what-s-story-5g. [Accessed 27 September 2017].

2 J. Penttinen, "Mobile Generations Explained," 20 November 2015. https://interferencetechnology.com/mobile-generations-explained. [Accessed 27 September 2017].

3 J. Penttinen, "Mobile Generations Explained," EMC Europe Guide, 20 November 2015.

4 D. Samberg, "Verizon sets roadmap to 5G technology in U.S.; Field trials to start in 2016," Verizon, September 2015.

5 M. Branda, "Qualcomm Research demonstrates robust mmWave design for 5G," Qualcomm Research, 15 November 2015.

6 J. Penttinen, "What's the story with 5g," Electronic Design, 2016. http://electronicdesign.com/communications/what-s-story-5g. [Accessed 04 July 2018].

7 IoTBlog, "Hands on with LoRa in North America," http://IoTBlog.org, 15 November 2015. https://iotblog.org/lora-hands-on. Accessed 31 July 2018.

8 S. Kinney, "100 US cities covered by Senet LoRa network for IoT," RCR Wireless News, 15 June 2016. Available: http://www.rcrwireless.com/20160615/internet-of-things/100-u-s-cities-covered-senet-lora-network-iot-tag17. [Accessed 31 07 2018].

9 LORIOT, "LORIOT aunches new LORAWAN server in the US," LORIOT, 01 April 2016. https://www.loriot.io/news/2016-04-01-loriot-launches-new-lorawan-server-in-the-us.html. [Accessed 31 July 2018].

10 Ingenu, "Press Releases," Ingenu, 2018. [Online]. Available: http://www.ingenu.com/category/press-releases. [Accessed 31 July 2018].

11 Tomás, J.P. "T-Mobile to launch NB-IoT network in major cities across the Netherlands," RCR Wireless News, 20 October 2016. http://www.rcrwireless.com/20161020/europe/t-mobile-launch-nb-iot-network-major-cities-across-netherlands-tag23. Accessed 31 July 2018.

12 Anderson, M. "Huawei to deploy NB-IoT to full market early 2017," The Stack, 31 October 2016. https://thestack.com/iot/2016/10/31/huawei-nb-iot-2017. [Accessed 31 July 2018].

13 M. Dano, "Vodafone's U.S. IoT chief: "Forget Sigfox and LoRa, NB-IoT 'makes a significant amount of sense," FierceWireless, 2016.

14 Huawei, "5G whitepaper, MWC Barcelona, 2016," Huawei, February 2016. http://www.huawei.com/minisite/4-5g/img/4.5GWhitepaper.pdf. Accessed 04 07 2018.

15 3GPP, "RAN approved REL-13 NB_IOT CRs (RAN#72):," 3GPP, 2017. http://www.3gpp.org/images/PDF/R13_IOT_rev3.pdf . Accessed 31 July 2018.

16 R. Wilson, "Europe goes for LoRa IoT network technology without UK," Electronics Weekly, 28 June 2016. http://www.electronicsweekly.com/news/europe-goes-for-iot-network-technology-without-uk-2016-06. [Accessed 29 July 2018].

17 IoT Blog, "Hands on with LoRa in North America," 15 November 2015. https://iotblog.org/lora-hands-on. [Accessed 29 July 2018].

18 K. Garvin, "Ingenu Refines Market Strategy to Focus on Enhanced RPMA Technology Development and IoT Platform as a Service Solution," Ingenu, 14 March 2018. http://www.ingenu.com/category/press-releases. Accessed 29 July 2018.

19 G. Vos, "What is LPWA for the IoT? Part 2: Standard vs. Proprietary Technologies," Sierra Wireless, 21 March 2018. https://www.sierrawireless.com/iot-blog/iot-blog/2016/08/lpwa_for_the_iot_part_2_standard_vs_proprietary_technologies. [Accessed 29 July 2018].

20 Sigfox, "Sigfox connect," Sigfox, 2018. www.sigfox.com. [Accessed 29 July 2018].

21 Disk91, "Make your IoT design on Sigfox or LoRa?," Disk91, 2016. https://www.disk91.com/2016/technology/internet-of-things-technology/make-your-iot-design-on-sigfox-or-lora. Accessed 29 July 2018.

22 Huawei, "NB-IoT enabling new business opportunities," Huawei, Shenzhen, 2015.

23 Tallinn Manual Process, https://ccdcoe.org/tallinn-manual.html. Accessed 13 November, 2018.

24 Mosbergen, D. "Stephen Hawking Says Artificial Intelligence 'Could Spell the End of the Human Race,'" *Huffington Post* (2 December, 2014), https://www.huffingtonpost.com/2014/12/02/stephen-hawking-ai-artificial-intelligence-dangers_n_6255338.html.

25 "cyber". Oxford Living Dictionaries. https://en.oxforddictionaries.com/definition/us/cyber. Accessed 13 November 2018.

4

Architecture

4.1 Overview

The 5G architecture has gone through a major facelift compared to any of the previous mobile communications generations. Many new concepts will be introduced, from which the most significant ones are the network virtualization and edge computing, whereas the radio interface will be built newly via more advanced modulation schemes and advanced multi-array intelligent antenna solutions.

The International Telecommunications Union (ITU) dictating the overall performance requirements that are finalized as of the end of 2017, the 3rd Generation Partnership Project (3GPP), defines the technical requirements for the evolution path from Global System for Mobile Communications (GSM), Universal Mobile Telecommunications System (UMTS)/high-speed packet access (HSPA) and Long-Term Evolution (LTE). The New Radio (NR) designed for 5G can support a variety of significantly more complex use cases and increased capacity required for complying with massive machine type communications, ultra-low latency and significantly increased data throughput.

The 5G network architecture can be characterized as a *service based*. It refers to the ability of the network, wherever suitable, to present the architecture elements as network functions (NFs) offering their specialized services via interfaces of a common framework to any network functions that are permitted to make use of these provided services. Network Repository Functions (NRFs) have a special role in this architecture, as they allow each network function to discover the services offered by other network functions. This architecture model provides benefits of the modern ways of virtualization and software technologies in the deployments.

Furthermore, 5G provides the possibility for mobile network operators (MNOs) to deploy *network slicing (NS)*. In the 5G context, NS refers to a set of 3GPP-defined features and functionalities forming a complete Public Land Mobile Network (PLMN), which provides services to a set of user equipment (UE). Network slicing provides the means to utilize in a controlled way the needed set of PLMN functionalities and services so that the resources of the network can be optimized per use case.

Continuing with the benefits of 5G, its data services provide more options for customization compared to previous mobile generations, which is beneficial in the fluent application utilization, as the 5G system architecture (SA) also contains a novel quality of service (QoS) model.

Another benefit of the 5G system (5GS) is the generalized design of the functionalities and a forward-compatible interface between the access network (AN) and core

5G Explained: Security and Deployment of Advanced Mobile Communications, First Edition.
Jyrki T. J. Penttinen.
© 2019 John Wiley & Sons Ltd. Published 2019 by John Wiley & Sons Ltd.

network, which allows the *common 5G core network (5GC)* to operate with different AN. In the first phase of 5G, as defined in the 3GPP Release 15, the interworking networks that are possible to be utilized together with the 5G are the Next Generation Radio Access Network (NG-RAN) as defined by 3GPP, and the 3GPP-defined untrusted Wireless Local Area Network (WLAN) access. As has been the case previously, also 5G will have its evolution path, and the Release 16 will bring along more functionalities as the work progresses.

The 3GPP Release 15, as defined by the System Architecture group, SA, describes the 5G system architecture via a set of features and functionality that are essential for deploying operational 5G systems in the field. The relevant technical specifications for the complete description are TS 23.501, TS 23.502, and TS 23.503 from which stage 2 includes the overall architecture model and principles. It also describes the evolved Multimedia Broadband (eMBB) data services, subscriber authentication, and service usage authorization, application support in general and via edge computing, support for Internet Protocol Multimedia Subsystem (IMS) with emergency and regulatory services, interworking with 3GPP's 4G networks, and user services with variety of access systems such as fixed network access and WLAN.

This chapter presents the 5G architecture with some of the most relevant scenarios in roaming and nonroaming environments.

The 5G system as defined by the 3GPP is a major enhancement for the previous mobile communication generations. Along with the considerably better performance, the complete philosophy of the network architecture gets a major facelift. The most remarkable difference between the old world and the NR and core networks is the virtualization of the functionality, which provides much wider set of use cases. Furthermore, the new architecture model provides a fluent modernization of the network and utilization of the cloud principles for processing and storing data.

4.2 Architecture

This section presents the default 5G architecture with new and reutilized elements, connected and cooperative networks, aggregated services, and key functions of evolved networks.

4.2.1 Architecture of the Previous 3GPP Generations

To understand the new aspects of the 5G system, let's have a brief look on the previous generations. Figure 4.1 summarizes the key elements that have been present in the mobile networks since the deployment and evolution of 2G, 3G, and 4G. Along with the 5G, both radio and core networks will be renewed with new elements and interfaces present. Figure 4.2 further summarizes the familiar interfaces within the 2G, 3G, and 4G networks.

4.2.2 Architecture Options for 5G

At present, there are various options for adding 5G network to the infrastructure depicted in Figure 4.2. 3GPP has considered the final architecture and the respective

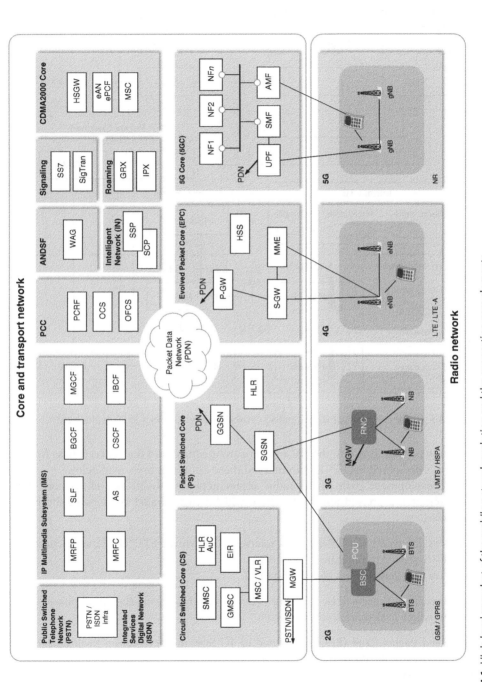

Figure 4.1 High-level snapshot of the mobile network evolution and the supporting core elements.

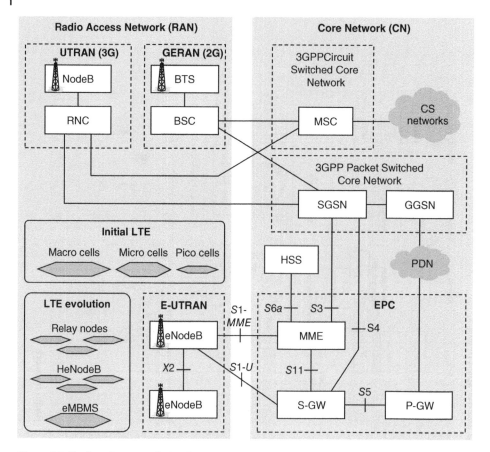

Figure 4.2 The key elements of 2G, 3G, and 4G network architecture prior to 5G deployment.

evolution plan leading into the NR and core architecture, and the concrete base for the options is presented in [1]. The summary of these options is presented below.

The development path of 5G radio access network (RAN) as well as core network requires a phased approach. There will be two parallel versions of the radio technologies:

- LTE as defined in 3GPP Release 15, referred to as evolved LTE (eLTE), as discussed in 3GPP System Architecture group SA2.
- Native fifth generation's New Radio (NR).

Furthermore, there will be two core network variants during the initial phase of 5G system (which is also referred to as 5GS):

- Evolved Packet Core (EPC) as defined in 4G System Architecture Evolution (SAE) technical specifications of 3GPP.
- Next Generation Core Network (NGCN).

The deployment strategies of these variants lead into 12 scenarios, as summarized in [1].

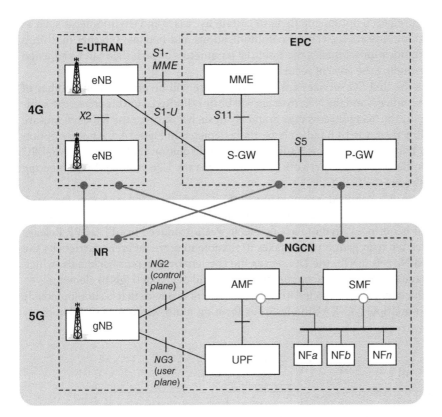

Figure 4.3 The high-level principle for the architecture of the LTE/EPC and 5G new radio (NR)/next-generation core network (NGCN). The 5G system provides possibility to virtualize the network functions (NF), which is presented in conceptual level in this format. When 5G is set to interwork via 4G EPC, the S-GW and P-GW are further subdivided into separate functional elements, each taking care of user and control planes.

The scenarios representing the standalone LTE/EPC and standalone NR/NGCN architectures are depicted in Figure 4.3.

As an additional solution in the interim phase of 4G EPC support for the initial 5G deployments, the Serving Gateway (S-GW) and Proxy Gateway (P-GW) functions have been broken into two parts: user plane (UP) and control plane (CP).

Figure 4.3 depicts some of the most important new 5G-specific elements:

- AMF refers to the Access and Mobility Management Function.
- UPF refers to the User Plane Function.
- SMF refers to the Session Management Function.
- NF refers to a network function. 5G network has a number of these network functions for specialized tasks, in a virtualized environment.

The philosophy of the 5G is remarkably different from the previous generations due to the virtualization of the network functions. This results to new, more optimized ways of utilizing the resources in a dynamic manner upon the need, via network slices. Network slice refers to a logical network that provides specific network capabilities and network characteristics. The MNO can have a number of these network slices that can be configured each optimizing them to certain use cases and environments such as broadband communications, critical communications and massive Internet of Things (mIoT) communications. Equally, the essential parameters of each slice can be adjusted separately, including the level of security.

As soon as the first 5G services are deployed, there will be only a small number of 5G-capable terminals, and the NR coverage will be only limited. For this phase, the solution is to deploy the 5G radio coverage gradually in such a way that the NR elements are connected directly to the LTE radio elements as depicted in Figure 4.4. Another option is to provide both interfaces New Radio-Long-Term Evolution (NR-LTE) and NR-EPC for relying on the legacy core network of 4G either via the LTE radio, or by connecting the 5G NR directly to 4G EPC.

The first set of technical specifications of 5G includes the concept for NSA (non-standalone) architecture, which will be the option for dual connectivity (DC). The DC itself is not new anymore as such as it was introduced in LTE 3GPP Release 12. Where carrier aggregation (CA) makes it possible to serve different carriers by the same evolved NodeB (eNB), the DC serves the carriers by different backhauls, which refers to separate eNB elements, or combined eNB and 5G NodeB (gNB) elements.

DC, as defined in [2] and depicted in Figure 4.5, refers to device that is simultaneously connected to multiple NG-RAN nodes, thus forming multi-Radio Access Technology

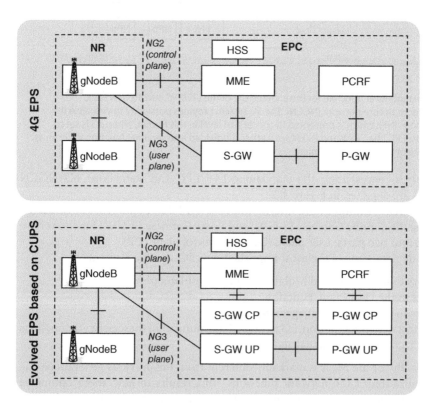

Figure 4.4 The 4G EPC develops further for better supporting 5G connectivity, by dividing the S-GW and P-GW functions into user and control planes.

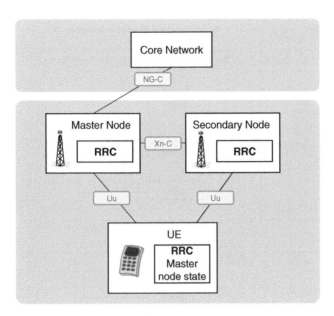

Figure 4.5 5G multi-DAT dual connectivity.

Dual Connectivity (MR-DC) with 5G Core network as defined in 3GPP TS 37.340. There are two possible scenarios for the DC:

- Next Generation Radio Access Network Evolved Universal Terrestrial Radio Access-New Radio Dual Connectivity (NGEN-DC). In this scenario, a device is connected to a Next Generation evolved NodeB (ng-eNB), which assumes the role of a master node (MN), and to a gNB, which acts as a secondary node (SN). The ng-eNB connects to the 5G Core, while the gNB connects to the ng-eNB via *Xn* interface.
- New Radio-Evolved Universal Terrestrial Radio Access Dual Connectivity (NE-DC). In this case, the device connects to a gNB, which assumes the role of a MN and an ng-eNB being a SN. The gNB connects to 5G Core and the ng-eNB connects to the gNB via *Xn* interface.

4.2.3 Gradual Deployment of 5G

For the fluent introduction of the 5G services, 3GPP has designed a phased approach for the deployment of the 5G. It means that the 5G operators may select the optimal network strategy along with the route toward full 5G services as they develop basing on practical equipment availability and further development of the 5G technical specifications as of Release 15 (referred to as the first phase of 5G), and Release 16 (second phase of 5G).

Figure 4.6 presents the 5G standalone scenario, which are called option 1 and 2 for 4G and 5G systems, respectively. Figure 4.6 also depicts the intermediate options paving the way for the fully native 5G standalone option 2:

- *Option 3*. The non-standalone scenario for the initial phase of the 5G deployments. The 5G terminal relies on both LTE and NR radio networks and solely on 4G core.

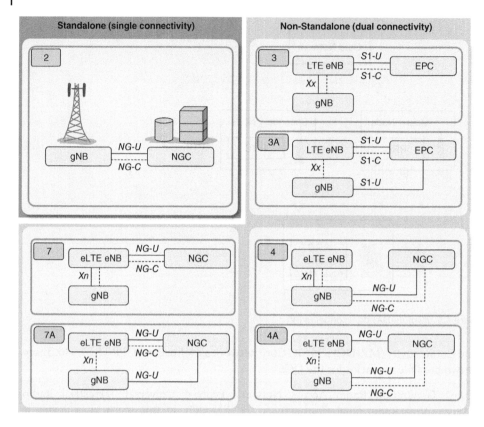

Figure 4.6 The most concrete 5G deployment options summarized.

- *Option 4.* The evolution path of the 5G can lead into the intermediate step of non-standalone, NR-assisted architecture that relies on the 5G core.
- *Option 5.* As soon as there is Release 15 LTE available, it can serve the further evolved LTE-Pro terminals while the very first NR serves the 5G-capable terminals separately. This option has been deprioritized.
- *Option 6.* The 5G terminal can be connected via booth 4G and 5G radio networks, while the 5G core takes care of the connectivity. This option provides the customers with the widest options to fully exploit the capabilities of the 5G, still providing the legacy connectivity via 4G. In this scenario, the 4G would take care of the major part of signaling while 5G radio delivers the data. This option can be compared, thus, with the principle of the carrier aggregation. This option has been deprioritized.
- *Option 7.* The 4G core can also take care of the 4G and 5G terminals in intermediate stage.

From these options, the ones summarized in Figure 4.6 have been selected as priority deployment variants.

To complete the first phase of 5G as defined in the 3GPP Release 15, the RAN 79 plenary meeting of 3GPP decided the phased approach in the following way:

- The options for 5G deployment scenarios are divided into three parts in the Release 15.
- The so-called late drop of the 5G NR was added into the Release 15 specifications on December 2018, about six months after the formal publication of the Release 15. This late drop refers to the architectural options of the NR that are not completed by September ASN.1 drop. Options 4 and 7 are part of this late drop.

The details of the selected options are defined in 3GPP TS 38.300 (Stage 2 radio architecture). The ng-eNB refers here to a node providing Evolved Universal Terrestrial Radio Access (E-UTRA) user plane (UP) and control plane protocol terminations toward the UE and connected via the *NG* interface to the 5GC. gNB, in turn, is a node providing NR user plane and control plane protocol terminations toward the UE and connected via the *NG* interface to the 5GC.

4.2.4   5G User and Control Plane

The 3GPP TS 38.401 defines the NR architecture and respective interfaces. One of the main differences in the 5G architecture compared to the previous ones is the logical separation of signaling and data transport networks. In fact, in addition to the cases within 5G infrastructure itself, this separation of the signaling and user data connections can be extended to cover the 5G and other access networks of previous generations so that the user data could go through 5G radio network, and the signaling load could be diverted via the 4G radio. This functionality can be understood as an extension to the carrier aggregation.

Another key aspect in 5G is that the 5G RAN and 5G core network functions are completely separated from transport functions. Thus, as an example, the addressing scheme of the 5G RANs and 5G core networks are not tied to the addressing schemes of transport functions. This means that the NG-RAN completely controls the mobility for an Radio Resource Control (RRC) connection.

For the NG-RAN interfaces, the specifications provide a fluent functional division across the interfaces as their options can be minimized. Furthermore, the interfaces are based on a logical model of the entity controlled through this interface, and one physical network element can implement multiple logical nodes.

This division of the user data and signaling traffic also means that the *Uu* and *NG* interfaces include both user plane and control plane protocols. The user plane protocols implement the packet data unit (PDU) session service carrying user data via the access stratum (AS) while the control plane protocols control the PDU sessions and the connection between the UE and the network, such as service requests, transmission resource control, and handover [3].

The 5G *user plane* is designed for transferring data via the PDU session resource service from one service access point (SAS) to another [3]. The high-level involvement of the 5G protocols is depicted in Figure 4.7. The respective RAN and core network protocols provide the PDU session resource service. The 5G radio protocols are defined in 3GPP TS 38.2*xx* and 38.3*xx* series, whereas the 5G core protocols are found in TS 38.41*x* series.

Figure 4.8 shows the high-level idea of the *control plane*, i.e. the principle of the division of the signaling protocol stacks of the 5G radio and core networks. For the control

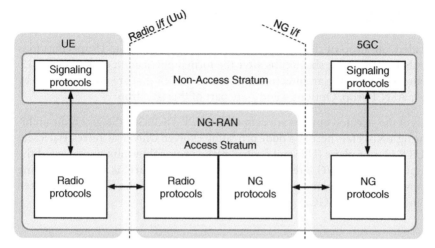

Figure 4.7 The principle of the 5G NG and Uu user plane [3].

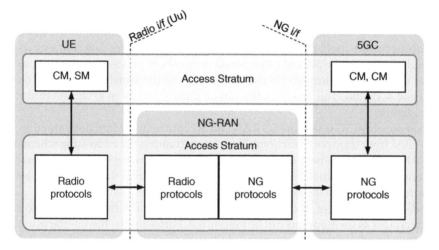

Figure 4.8 The control plane for Uu and NG.

plane, the protocols are defined in the same technical specification series 38.2*xx*, 38.3*xx*, and 38.41*x* as for the user plane. The radio and the NG protocols contain a mechanism to transparently transfer non-access stratum (NAS) messages. Please note that the control plane protocols shown in Figure 4.8 are presented as examples. CM refers to connection management and SM to session management.

4.2.5 NG-RAN Interfaces

The 5G network architecture can be depicted by using reference point representation or service-based interface representation. Figure 4.9 depicts the overall architecture and interfaces of the 3GPP 5G network in the traditional reference point format. As can be seen in the figure, the new terminology for the 5G core network is 5GC, whereas the NR system is referred to as NG-RAN.

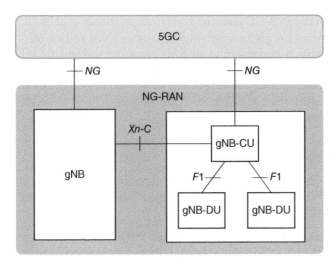

Figure 4.9 The overall 5G architecture.

The logical interfaces of the 5G architecture are referred to as *NG*, *Xn*, and *F1*. The complete description of these interfaces can be found in the following specifications:

- 3GPP TS 38.410 defines the general aspects and principles of the *NG* interface.
- 3GPP TS 38.420 defines the general aspects and principles of the *Xn* interface.
- 3GPP TS 38.470 defines the general aspects and principles of the *F1* interface.

The 5GC includes still the same type of network elements as is defined in the EPC of the LTE prior to the Release 15, i.e. Mobility Management Entity (MME) for signaling, S-GW, P-GW, and supporting elements such as for the policy rules, added by various functional 5G elements.

For the radio system, the base station elements are referred to as gNB, or next-generation NodeB. The *NG* interface interconnects the gNB elements of the NG-RAN and the 5GC. The *Xn* network interface is used between NG-RAN nodes.

In 5G, the gNB elements can support FDD (frequency division duplex) and TDD (time division duplex) modes, as well as dual-mode operation. The gNB elements are interconnected via the *Xn-C* interface.

The NG-RAN is defined via Radio Network Layer (RNL) and Transport Network Layer (TNL). The NG-RAN interfaces are defined as part of the RNL, whereas the TNL provides services for user plane transport and signaling transport. TNL protocol and the functionality are specified for the NG-RAN interfaces.

Each gNB may consist of a single 5G NodeB-Centralized Unit (gNB-CU) and a set of 5G NodeB-Distributed Units (gNB-DU). The connection within a gNB, between the gNB-CU and gNB-DU, takes place via the *F1* interface – either so that a single gNB-DU is connected to only one gNB-CU, or depending on the implementation, a single gNB-DU may also be connected to multiple gNB-CU elements. The connections from outside of the gNB element form other gNB elements and from the 5GC terminate always to the gNB-CU element.

Also, in case of the Evolved Universal Terrestrial Radio Access-New Radio Dual Connectivity (EN-DC), the *S1-U* and *X2-C* interfaces for a gNB that has the gNB-CU and gNB-DU elements terminate in the gNB-CU element. Nevertheless, the set of gNB-CU and one or more gNB-DU elements per gNB are only visible to other gNBs and the 5GC as a gNB. Furthermore, each gNB is connected to all AMF elements within an AMF Region in NG-Flex configuration. The AMF region is defined in 3GPP TS 23.501 [4]. It consists of one or more AMF sets. The AMF set, in turn, consists of some AMF elements that serve a given area and network slice. Multiple AMF sets may be defined per AMF region and network slice or slices.

4.2.6   Functions of gNB and ng-eNB

The functions of the native 5G gNB and evolved 4G ng-eNB supporting connectivity to 5G system are defined in 3GPP TS 38.300. It states that both elements host the following functions:

- Internet protocol (IP) header compression, data encryption, and integrity protection
- Selection of an AMF during the UE attachment procedure in such case when no routing to an AMF can be determined from the information provided by the UE
- Routing of user plane data toward UPF and control plane information toward AMF
- Connection setup and release
- Scheduling and transmission of paging messages and system broadcast information provided by the AMF or O&M
- Measurement and respective reporting configuration to be used for the tasks related to mobility and scheduling
- Uplink packet marking at the uplink transport level
- Session management
- Functionality related to network slicing
- QoS flow management and mapping it to data radio bearers (DRBs)
- Support of devices in the RRC_INACTIVE state
- Distribution function for NAS messages
- RAN sharing
- DC
- Tight interworking between NR and E-UTRA
- Radio Resource Management (RRM)

The RRM function includes radio bearer and admission control, connection mobility control, as well as scheduling, which refers to the dynamic allocation of uplink and downlink resources for devices.

4.2.7   5G System Architecture

The 3GPP technical specification TS 23.501 describes the 5G system architecture. This section presents some of the key architecture models referencing the technical specification.

The interaction between the 5G network functions can be represented in two ways; via the traditional reference point model or service-based model. The service-based representation refers to the network functions within the control plane enabling other network functions to access their services. The reference point representation, in turn, depicts the interaction between the NF services in the network functions.

The NFs have thus both functional behavior and interface. Each NF can be deployed as a dedicated hardware forming the network element, or as a software instance that is relying on a hardware. Furthermore, the NF can also be implemented as a virtualized function instantiated on a relevant platform such as in a cloud-based infrastructure.

4.2.8   Network Functions (NF)

The 5G architecture consists of RAN basing on the *NR* interface, and NGC (next-generation core) that has renewed network elements. The access into 5G network can also take place via non-3GPP access network (AN). AN refers thus to a general radio base station, including non-3GPP access such as Wi-Fi.

The accessing device is the UE, which is formed by the mobile terminal (MT) and a tamper-resistant secure element (SE), which may have a form of "traditional" UICC (universal integrated circuit card), commonly known as SIM card (subscriber identity module), or its evolved variant such as embedded universal integrated circuit card (eUICC). If the UE connects to a data network (DN), it can be, e.g. operator services, Internet access, or third-party services.

The 5G system (5GS) architecture consists of a set of network functions (NF), of which each has their own defined task. The 3GPP TS 23.501 includes the functional description of these network functions that are listed in Table 4.1 and detailed in the next sections.

It can be seen, based on Table 4.1, that some of the functional elements represent a major change to the LTE while others are merely a more "peaceful" evolution by adding performance and functionality to support the new network. The most significant changes are related to S-GW and P-GW elements, referred to as S/P-GW in the collocated form, which now is divided into a user plane via the UPF and into control plane via the SMF. Equally, the evolution of the Authentication, Authorization and Accounting-Home Subscription Server (AAA-HSS) results in renewed Authentication Server Function-Unified Data Management (AUSF-UDM), and the data storage is now, in the 5G era, taken care of by the Structured Data Storage Network Function (SDSF) and the Unified Data Repository (UDR) (Figure 4.10).

4.2.9   Architecture of 5G

The great difference between the previous generations and the new 5G is that the system architecture of the latter can be service based. That refers to the possibility, wherever suitable, for defining the architecture elements as network functions (NF) offering their specially designed services via common framework interfaces to any other network function if they have permission to do rely on the respective, offered services. So, as in 4G EPC, there are defined protocol and reference points for each entity, like S-GW, P-GW, and MME, whereas in the 5G, these protocol and reference points are defined for all the available NFs. In order to understand if the NF has certain rights, the NRFs are defined. They allow network functions to discover the services offered by other network functions.

The service-based architecture model provides highly modular and reusable approach for the 5G network deployments. It thus allows MNOs to deploy the 5G gradually, benefiting from the latest advancements of the virtualization concept. The service-based architecture contains the core network functions by interconnecting them simply

Table 4.1 Network functions (NF) and elements of the 3GPP 5G system.

NF	Description	Mapping with 4G
5G-EIR	Equipment identity register	Evolution of LTE EIR
AF	Application function	Application layer of 5G, comparable with the LTE AS and gsmSCF
AMF	Access and mobility management function	Replaces the LTE MME
AUSF	Authentication server function	Replaces the LTE MME/AAA
NEF	Network exposure function	Evolution of SCEF and application programming interfaces (APIs) layer
NRF	NF repository function	Part of the evolution of DNS
NSSF	Network slice selection function	New function for 5G-specific network slicing concept
PCF	Policy control function	Evolution of the LTE PCRF (policy and charging enforcement function)
SEPP	Security edge protection proxy	New element for securely interconnecting 5G networks
SMF	Session management function	Replaces together with the 5G UPF the LTE S-GW and P-GW
UDM	Unified data management	Evolution of HSS and UDR
UDR	Unified data repository	Evolution of the LTE SDS (structured data storage)
UDSF	Unstructured data storage function	The function comparable with LTE SDSF (structured data storage network function)
UPF	User plane function	Replaces together with the SMF the LTE S-GW and P-GW

with the rest of the system. Figure 4.11 presents the visual model for depicting the service-based architecture, combined with the reference point-based architecture elements. The latter shows the interactions between network functions to inform about the system level functionality, as well as inter-PLMN interconnection via different network functions.

4.2.9.1 Roaming and Non-Roaming 5G Architecture

Figure 4.11 shows the reference architecture of the 5G network basing on the service-based interfaces and Figure 4.12 shows the alternative format of the nonroaming network architecture basing on the reference point representation. In addition to the presented elements, the full architecture also contains the rest of the network functions listed in Table 4.1. All the presented network functions are capable of interacting with the UDSF (Unstructured Data Storage Function), UDR, NEF (Network Exposure Function) and NRF (NF Repository Function) when appropriate.

In Figure 4.11, the terms described in Table 4.1, the tasks of the presented elements are the following; the more detailed description of the elements cab be found in 3GPP TS 23.501 [4].

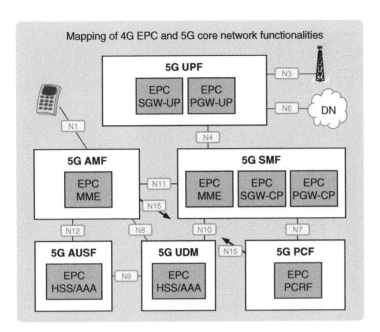

Figure 4.10 The comparison of functional elements of the 4G and 5G. Please note that in this figure, the EPC elements are shown only for indicating the similarities with the respective 5G elements, but this is not the actual interworking architecture.

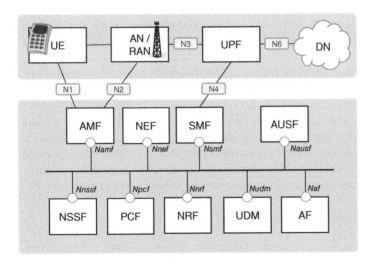

Figure 4.11 Non-roaming 5G system architecture, presented via service-based interfaces.

- UE represents consumer device like smart device, or Internet of Things (IoT) device such as machine-to-machine (M2M) device. It comprises the device (HW and SW) and the tamper-resistant secure element such as UICC.
- AN/RAN is the 5G NR access network or other access network, including Wi-Fi access.

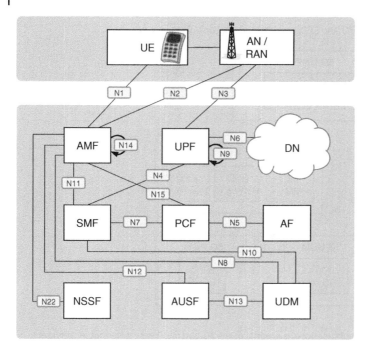

Figure 4.12 Reference point presentation of the 5G system architecture for the non-roaming case [4].

- UPF is the user plane function. UPF interacts between the 5G network and external data network, supporting PDU Session Anchor functionality. Its role is thus to act as an external PDU Session point of interconnect to the data network. It does not provide any services in the Release 15 3GPP technical specifications, but it can consume services provided by 5GC Control Plane NFs. UPF Service Area refers to an area within which PDU Session associated with the UPF can be served by AN and RAN nodes via an *N3* interface between the access network and the UPF without need to add a new UPF in between or to remove or reallocate the UPF. Uplink classifier (UL-CL) is UPF functionality that aims at diverting uplink traffic, based on filter rules provided by SMF, toward the data network. Please note that a single UE may establish multiple PDU Sessions to the same data network and served by different UPF terminating *N6*.

The functional elements of the Figure 4.11 have the following tasks:

- AMF is access and mobility management function. It has multitude of functions, including registration, connection, reachability and mobility management. It also participates in access authentication and authorization, providing security anchor functionality (SEAF). It interacts with the Authentication Server Function (AUSF) and the UE, receives the intermediate key from the UE authentication process. If universal subscriber identity module (USIM-based authentication takes place, it retrieves the security material from the AUSF. It also performs security context management (SCM) by receiving a key from the SEAF to derive access network-specific keys. It also stores the UE capability list, performs load balancing, and interacts with SMF.

- SMF takes care of session management and roaming functionality, among a multitude of tasks. It establishes, modifies, and releases sessions. It manages UE's IP addresses and Dynamic Host Control Protocol (DHCP) v4/v6 server and client functions.
- NEF takes care of the exposure of capabilities and events, among many other tasks. It stores and retrieves information as structured data. It also makes secure provision of information from external application to 3GPP network.
- AUSF supports authentication procedures as specified by SA WG3.
- NSSF is Network Slice Selection Function. selects the set of NS instances serving the UE, determines the allowed NSSAI, configured NSSAI, and the AMF set to be used to serve the UE or a list of candidate AMF(s).
- Policy Control Function (PCF) supports unified policy framework to govern network behavior and provides policy rules to control plane functions.
- NRF supports service discovery function and maintains the NF profile of available NF instances and their supported services.
- Unified data management (UDM) generates 3GPP Authentication and Key Agreement (AKA) authentication credentials. It handles user identification by storing and managing subscription permanent identifiers (SUPIs) per 5G subscriber. It also manages Short Message Service (SMS).
- AF is application function, which interacts with the 3GPP core network to provide services such as application influence on traffic routing, access to NEF, and interaction with the policy framework.

The other participating elements not shown in this figure include the following, with the respective tasks:

- UDSF is Unified Data Repository, which supports, e.g. storage and retrieval of subscription data by the UDM, storage and retrieval of policy data by the PCF, and storage and retrieval of structured data by the NEF. UDSF is located in the same PLMN as the NF in such a way that service consumers store in and retrieve data from it via *Nudr* (intra-PLMN interface). The UDR may collocate with UDSF.
- UDR is an optional function for storage and retrieval of information as unstructured data by any NF.

There may also be other functional elements such as the Short Message Service Function (SMSF), which supports SMS over NAS, 5G-Equipment Identity Register (5G-EIR) to check the status of PEI blacklisting, Location Management Function (LMF) for UEs location determination and measurements of UE while Network Data Analytics Function (NWDAF) represents operator managed network analytics logical function.

Figure 4.13 presents another variant of the Figure 4.11, now with a roaming case.

Figure 4.13 shows the service-based architecture for a roaming basing on local breakout (LBO). In LBO scenario the SMF and all UPF(s) involved by the PDU Session are under control of the visited public land mobile network (VPLMN). In this scenario, the UE, which is roaming in visited network (VPLMN), establishes connection to the data network of the VPLMN while the home public land mobile network (HPLMN) enables the connectivity basing on the user's subscription information (UDM), subscriber authentication (AUSF) and policies (PCF) for this specific UE. The interworking between HPLMN and VPLMN is protected by home security edge protection proxy (hSEPP) and visited network's security proxy (vSEPP). The Security Edge Protection Proxy (SEPP) is nontransparent proxy and supports message filtering and policing on inter-PLMN control plane interfaces, and topology hiding.

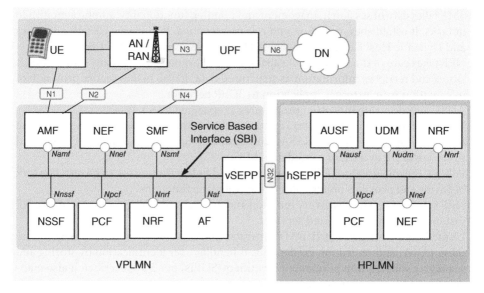

Figure 4.13 An example of the 5G architecture in a roaming case, presented using service-based interfaces.

In this example, the visited network provides functions for the network slice selection (NSSF), network access control and mobility management (AMF), data service management (SMF) and application functions (AF). The same principles for the separate user and control planes managed by the user plane (UPF) is applied in 5G as it has been defined already in 4G.

Continuing with the presentation of the 5G architectural examples, Figure 4.14 shows the reference point format of the non-roaming architecture for UEs concurrently accessing local and central data networks using multiple PDU sessions with two SMF elements are selected for the two different PDU sessions.

Figure 4.15 shows the principle of the non-roaming architecture when concurrent access to two data networks is occurring within a single PDU session.

To finalize the architectural examples, Figure 4.16 shows the non-roaming architecture for NEF. In this example, the 3GPP interface represents the set of southbound interfaces between NEF and 5G Core Network Functions. These interfaces include, e.g. *N29* for Network Exposure Function–Session Management Function (NEF–SMF), and *N30* for Network Exposure Function–Policy Control Function (NEF–PCF), among other interfaces.

4.2.9.2 Interworking with Non-3GPP Networks

As an example of the non-3GPP access architecture scenarios, Figure 4.17 presents the non-roaming architecture for 5G core network via non-3GPP access network. The UE can be connected simultaneously to the 5G core via 3GPP and non-3GPP access networks in which case the single AMF is communicating via two *N2* interfaces.

Nevertheless, the two *N3* instances presented in Figure 4.17 can apply to a single or different UPF elements when separate PDU sessions are active based on 3GPP access and non-3GPP access networks.

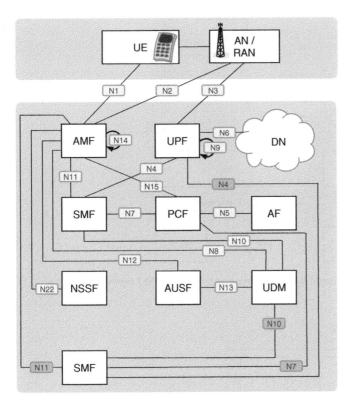

Figure 4.14 Reference point format of non-roaming 5G system architecture for multiple PDU sessions. The colored reference points differ from the architecture presented in Figure 4.12.

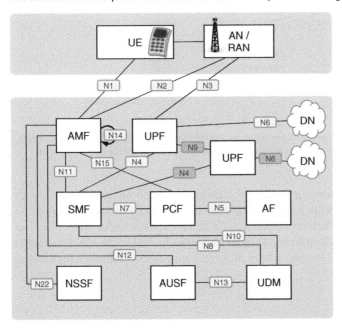

Figure 4.15 Reference point format of non-roaming 5G system architecture for concurrent access to two data networks according to the single PDU session option. The colored reference points differ from the architecture presented in Figure 4.12.

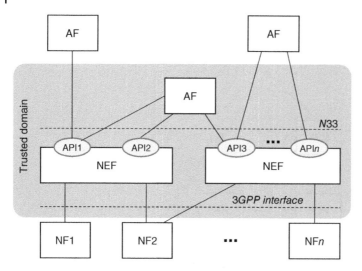

Figure 4.16 Reference point format of the non-roaming architecture for NEF (network exposure function).

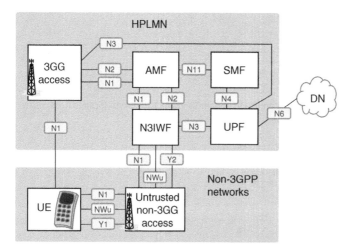

Figure 4.17 5G core network architecture for non-roaming via non-3GPP access. Please note that this figure exposes only the direct connections for the access networks. The rest of the architecture is presented in Figures 4.12 and 4.14.

This section ends by presenting the 5G architecture scenario for interworking with the 4G core network, i.e. EPC in non-roaming environment as depicted in Figure 4.18.

In Figure 4.18, the interworking between 4G and 5G core networks takes place via optional *N26* interface via 4G-specific MME and 5G-specific AMF. The *N26* interface is able to support the essential functionalities of the 4G-specific *S10* interface for making the interworking possible.

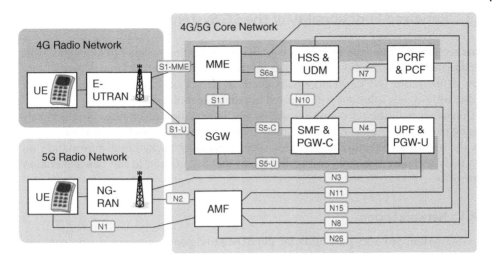

Figure 4.18 Non-roaming architecture for interworking between 5G system (5G new radio and 5G core) and 4G system (EPC and E-UTRAN). The 4G-specific elements are highlighted by darker gray color.

The PCF and Policy and Charging Enforcement Function (PCRF), SMF&PGW-C, and UPF&PGW-U are optional elements for the interworking between 4G and 5G systems and are able to serve the networks and UEs based on their supported set of capabilities. The UEs that are not able to support 4G–5G interworking are served by the elements of the respective 4G or 5G native elements, which are P-GW and PCRF for the 4G, and SMF, UPF, and PCF for 5G.

As dictated by the 3GPP TS 33.501, another UPF can be deployed between the NG-RAN and the UPF&PGW-U via additional interface *N9*.

In addition to the figures presented in this chapter, the 5G architecture cases contain various other scenarios as defined in TS 23.501 and other relevant technical 3GPP specifications. These cases include, e.g. the following components:

- Interworking between 5GC via non-3GPP access and E-UTRAN connected to EPC
- LBO roaming architecture for non-3GPP accesses, non-3GPP Interworking Function (N3IWF) in same PLMN as 3GPP access
- Home-routed roaming architecture for non-3GPP accesses, N3IWF in same PLMN as 3GPP access
- LBO roaming architecture for non-3GPP accesses, N3IWF in different PLMN from 3GPP access
- Home-routed roaming architecture for non-3GPP accesses, N3IWF in different PLMN from 3GPP access
- Network Analytics architecture

Please refer to the 3GPP TS 23.501 for more details on these cases.

4.3   Renewed Functionality of the 5G System

This section describes the new aspects of 5G by presenting insights to key items. This section includes description and analysis for the SW defined networking, network function virtualization (NFV), small cells, evolved HetNet, cloud-based functions, cognitive radio, white-space solutions, and evolved broadcast techniques.

Convergence of the telecommunications and information technology is remarkably taking place in 5G and is one of the main drivers of the changes that need to happen on the overall architecture of 5G. This section provides background information on the main technical concepts and technologies that are utilized in the new generation networks.

4.3.1   Network Slicing

The 5G system provides possibility for the MNO to deploy a network slicing concept. The network slice is a logical network that is deployed to serve a defined business purpose or customer, and it contains all the required network resources that are configured as a set to serve that specific purpose.

The network slice is an enabler for the services. The network slice can be created, modified and deleted via the respective network management functions. The network slice is thus a provider's managed, logical network, and it can be deployed technically in any network type including mobile and fixed networks.

Furthermore, the underlying resources can be physical or virtual, and they may even integrate services from different providers which may be beneficial for, e.g. roaming scenarios. As the network slicing concept provides much more flexibility and optimized utilization of the physical network resources, it is adapted to the 3GPP 5G system, too [5].

The network slicing concepts are beneficial for the MNOs, providing the possibility to deploy merely those functions that are required to support certain customers and market segments. Furthermore, deployment of additional resources is not needed to offer these planned functionalities, which has positive impact in terms of reduced expenses. Another benefit is the fast deployment and time to market [6].

As indicated in Ref. [5] and depicted in Figure 4.19, the network slicing concept consists of three layers – the service instance layer, network slice instance layer, and resource layer.

The network slice is in fact a set of network functions, which, in turn, represents a functional block. Each function can be provided by independent vendor providing wider grade of independency between infra vendors.

The functional block is located within a network infrastructure with external interfaces. MNO can configure and customize the network functions and thus create in a dynamic way optimized networks for different use segments. Figures 4.20 and 4.21 depict an example of the principle of the slice mapping between the NR and 5G core.

As can be observed in Figure 4.21, there are access network slices consisting of both radio and fixed network access types and core network slices. There also is a selection function in between to map these slices, which forms the complete slice via these components. The device communication with the access slices can be, e.g. smart phones and IoT devices. Each core network slice is optimized for specific service type such as mIoT, Critical Communications, MVNO, and eMBB services; or focusing even more, the slice

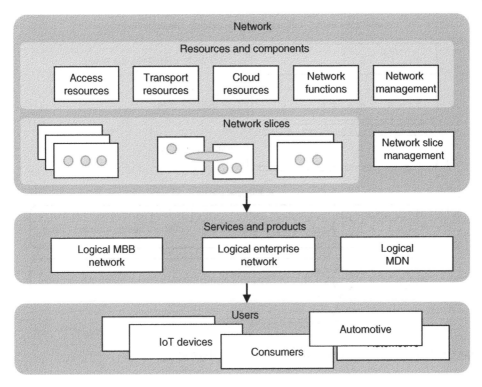

Figure 4.19 The principle of network slicing.

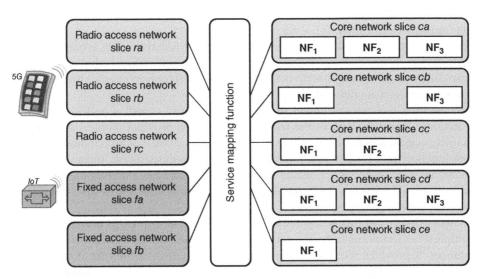

Figure 4.20 An example of the mapping of the network slicing components of the radio access and core network.

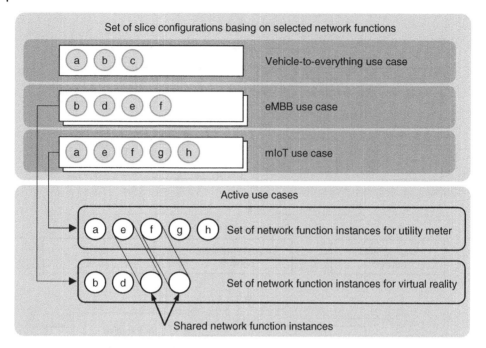

Figure 4.21 The principle of network slices with conceptual examples.

can be tailored for, e.g. a factory for remote control functions, for a virtual reality service, or a utility firm. Each slice contains the respective services such as the special MVNO features and charging principles for the MVNO slice, or low rate and energy efficient data transmission for the mIoT slice.

A single core network slice contains a set of network functions. The actual mapping of the device, access, and core network slices may be using 1:1:1 or 1:M:N model. This means that a single UE may use more than one slices, and each access slice may connect to more than one core network slice.

The network dedicates resources to run these functions, as well as policies and configurations. This is the task of the resource layer which provides physical and logical resources required to run each end-user and business service. For each moment, a network function instance (NF instance) takes care of the dedicated task, and a set of these NF instances makes the functionality for that specific use case. Different use cases use different sets of NF instances, but some of these NF instances can also be shared between use cases as depicted in Figure 4.21.

A network slice instance is selected by authorized UEs, allowing UEs to associate with multiple slices simultaneously. The network instance is security isolated from other slice instances. It provides access to common network functions of the core network and access network enabling network slicing. The network slice instance also supports roaming scenarios.

In summary, network slicing provides the MNO with the possibility to deploy multiple, independent PLMNs. Each of these PLMNs can be customized by instantiating only the features, capabilities, and services that are needed for the subset of the served UEs.

The requirements for the network slice selection are defined in 3GPP TR 23.799, which shows examples of the procedures. There are three types of network slicing scenarios identified, which are referred to as groups A, B, and C, to support more than one network slices per device:

- Group A refers to situations where the device is consuming services from multiple network slices and different core network instances, which are logically separated. Each network slice serves the UE independently as for the subscription and mobility management. The drawback of this scenario is the increased signaling in core and radio interface, although the isolation is probably easiest to achieve.
- Group B is a combination of shared network functions among slices while other functions are in individual slices.
- Group C refers to a scenario where the control plane is common amongst slices, and individual network slices handle the user plane (Figure 4.22).

4.3.2 Network Function Virtualization

Network slicing is based on the idea that results in the *virtualization* of network functions, which can be provided in software rather than in special-purpose hardware as in today's generations of networks. The NFV refers to a principle of separating network functions from the hardware they run on by using virtual hardware abstraction.

Current network technology is based on network functions that are implemented as a combination of vendor-specific software and hardware. These network functions are

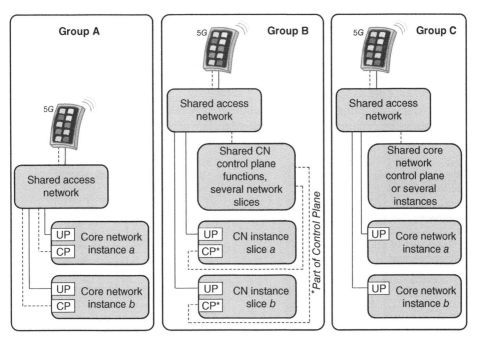

Figure 4.22 The groups A, B, and C to support multiple network slices per device, as interpreted from the 3GPP TR 23.799.

statically chained or connected to achieve the overall functionality. This static combination of network functions is called *Network Function Forwarding Graph,* or NF set. In 5G, as defined by 3GPP, these forwarding graphs will be composed dynamically, which makes it possible to address the current demands (Figure 4.23).

The NFV concept provides the possibility to decouple software from hardware, which allows a smooth evolution of both SW and HW independently. Furthermore, the flexible and automated NF SW instantiation deployment leverages available cloud and network technologies. As an additional benefit of the NFV, the dynamic operation and management of NF sets enhances the scaling of the network performance.

4.3.3 Open Source

Another trend in the overall information and communication technology (ICT) industry is open source. The http://OpenBTS.org is an open source software project. According to the statements of this organization, the open source concept is dedicated to revolutionizing mobile networks by substituting legacy telecommunications protocols and traditionally complex, proprietary hardware systems with IP and a flexible software architecture. This architecture is open to innovation by anybody, allowing the development of new applications and services and dramatically simplifying the setting up and operation of a mobile network.

It should be noted that in addition to a variety of benefits of the open source, as the source code of network components is now available to anybody easily, it may also be more susceptible to fraud, which must be addressed in standardization, productization, and deployment of 5G.

4.3.4 Mobile-Edge Computing

Driven by the requirement for ultra-low latency and high bandwidth, both network functions and content must move closer to the subscriber, i.e. as far toward the edge of the radio network as feasible.

Mobile-edge computing (MEC) allows content, services, and applications to be accelerated, increasing responsiveness from the edge. Furthermore, the mobile subscriber's

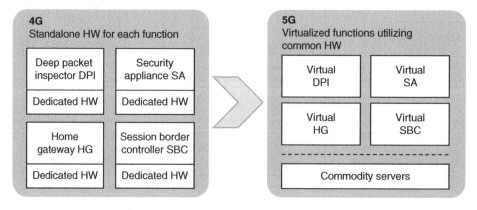

Figure 4.23 The principle of NFV concept.

experience can be enriched through efficient network and service operations, based on insight into the radio and network conditions. Those applications are implemented as software-only entities running on top of a virtualization infrastructure.

4.3.5 All-IP Technology

In the increasing demand of enabling new business models, 5G also requires reduced costs of the utilization and offering of the services.

NFV and software-defined networking (SDN) are technologies that help to reduce infrastructure and management expenses significantly. However, these mechanisms imply that communication between network entities is migrating from today's, in many cases, proprietary protocols to standard IP-based mechanisms, such as IPsec.

4.3.6 Service-Based Architecture

In 5G, as defined by 3GPP, the service-based architecture is defined between the control plane network functions of the core network. In 5G, the network functions can store their contexts in data storage functions (DSFs) and makes it possible to separate the data storage as for the UE, access network, and the AMF. So, there is no more such tight binding of the elements as in previous generations, which makes the functionality much more flexible and increases the performance by optimizing the resource utilization. The previous architecture models did the UE-specific transport association, which resulted in more complex tasks to change the UE's serving node (which is comparable to a 5G AMF). The 5G functionality notably simplifies the changing the AMF instance that serves a UE. The new architectural model also supports enhanced AMF resilience and load balancing. This is a result of the AMF functionality, which allows a set of AMFs within the same network slice to handle procedures of any UE.

4.4 Supporting Solutions for 5G

Apart from the architectural and functional 3GPP specifications, there are solutions under construction optimizing further the performance and capacity of the 5G networks. These include adaptive antennas and a further developed self-optimizing network (SON) concept.

4.4.1 Adaptives

The 5G has been developed and different concepts and technologies have been tested actively. The radio and core technologies have been upgraded, and there are novelty technologies expected to appear during the evolution path of the systems after Release 15 and 16 specifications. The demanding performance requirements of 5G can indeed be satisfied with a set of many different type of technologies that are orchestrated for optimizing the user experience.

One of the enabling technologies that has been under development for some time is the active antenna system (AAS). In 5G, this technology is expected to be further developed and deployed.

4.4.2 SON

Another technology familiar from the previous systems is the SON concept. It is a constantly evolving enabler that will also be in useful in 5G.

One of the related scientific field is the machine learning, which could assist in the optimal adjustment and fault recovery of the 5G networks. The SON is defined by 3GPP in the TS 38.300 and refers to self-configuration and self-optimization. This work item was finalized in June 2018 to be included into 5G as of Release 15. The specification includes definitions for the UE support for SON, self-configuration by dynamic configuration of the *NG-C* interface, dynamic configuration of the *Xn* interface, automatic neighbor cell relation (NCR) function. The latter is important, as it is designed to ease the task of the operator from the manually managing NCRs. There will thus be definitions for intra-system–intra NR and intra-system–intra E-UTRA automatic NCR function, and intra-system–inter RAT and inter-system automatic NCR function, among various other functions the SON can provide.

4.4.3 QoS in 5G

The 5G services are much more flexible than in previous generations. Much of this is due to the new QoS model of the 5G system architecture. Figure 4.24 depicts the principle of the 5G QoS as interpreted from 3GPP [7].

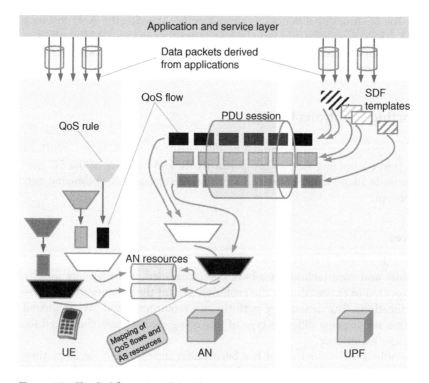

Figure 4.24 The QoS flow principle in 5G.

The 5G QoS makes it possible to differentiate the data services to cope with performance requirements of variety of applications while sharing the limited radio resources in an optimal way. Another aspect is that the 5G QoS supports the different requirements of access technologies that can be used for 5G connectivity. The access network can be even fixed that has very different requirements from the radio access, such as QoS support without additional signaling. One way to handle this is to deploy standardized packet marking, which indicates to the QoS-related functions on the desired QoS level without the extra signaling. 5G also supports symmetric QoS differentiation for downlink and uplink minimizing the control plan by adopting reflective QoS model.

The 5G supports flexible deployment of application functions, which is beneficial for, e.g. edge computing. 5G makes it possible to deploy QoS via a set of session and service continuity modes (SSC). In this concept, the SSC 1 is previously defined mode. It ensures that the IP anchor remains stable supporting applications and maintaining the link for UE in location updates. The new modes SSC 2 and SSC 3 make it possible to relocate the IP anchor. There are two new options, so-called "break-before-make" mode (SSC 2), and "make-before-break" (SSC 3). These enable applications to influence selection of suitable data service characteristics and SSC mode. Figure 4.25 depicts the principle of these modes.

In addition to the above-mentioned methods, some of the QoS-related functionalities 5G offers are UL-CL and branching point. As stated in Ref. [8], UL-CL is a functionality supported by a UPF that aims at diverting traffic locally to local data networks based on traffic matching filters applied to the UE traffic. The branching point refers to the functionality of UPF, which is the generalized logical data plane function with context

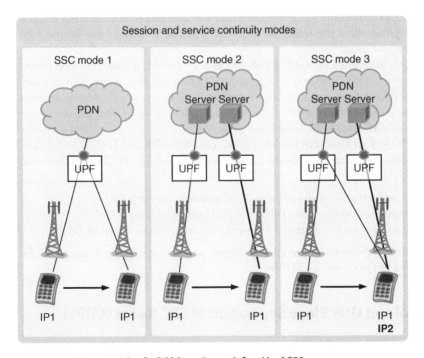

Figure 4.25 The principle of 5G SSC modes as defined by 3GPP.

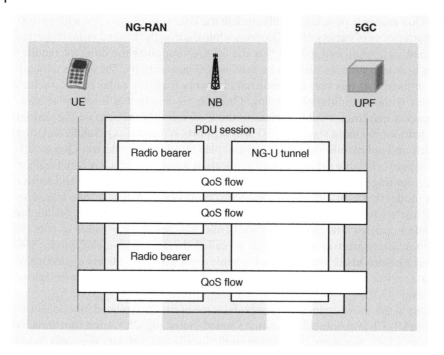

Figure 4.26 QoS flows in 5G.

of the UE PDU session. UPFs can play many roles, such as being a flow classifier basing on the UL-CL, a PDU session anchoring point, or a branching point. Both the UL-CL and branching point allow the injection of traffic selectively to and from application functions on the user plane links.

The QoS in 5G, as defined in 3GPP TS 38.300 and TS 23.501, is based on QoS flows. This model supports QoS flows requiring guaranteed flow bit rate, referred to as guaranteed bit rate (GBR) QoS flows. It also supports QoS flows, which do not require guaranteed flow bit rate, which is referred to as non-GBR QoS flows. This provides QoS differentiation at NAS level's PDU session.

The NG-RAN QoS architecture for 5G NR–5G core as well as for E-UTRA–5G core network is depicted in Figure 4.26. Some of the key principles for the 5G QoS flows are the following:

- 5G core network establishes one or more PDU sessions for each UE.
- NG-RAN establishes one or more DRBs per PDU session for each UE.
- NG-RAN maps packets belonging to different PDU sessions to different DRBs.

As a result, NG-RAN and 5G core are capable of jointly ensuring the desired QoS by mapping packets to QoS flows and DRBs.

4.5   Control and User Plane Separation of EPC Nodes (CUPS)

The 3GPP-defined 5G system has a variety of new concepts. Some of these are control plane and user plane split, network slicing, and service-based architecture. Although all of these are essential building blocks for complying with the strict requirements of radio section of International Telecommunication Union (ITU-R) IMT-2020, network slicing is a key enabler supporting multiple different use cases and instantiations of same functionality.

The 5G mobile broadband (MBB) use case is an evolution of the 4G broadband connectivity. The difference is that this is enabled via service-based architecture in 5G era. To take advantage of technical options and market needs, an interim solution provided by 3GPP is the support of diverging architectures for 5G services.

As indicated in [5], the 5G core standardization would ideally define a functional architecture where implemented technologies may evolve and can be replaced when time is adequate to do so. The support of multivendor environment is one of the important principles for achieving this goal and enhancing independently the user and control plane functionalities. The latter provides possibility for flexible deployment, allowing variable network configurations via network slices.

In this transition phase, the original 4G EPC will thus change from the signaling point of view as depicted earlier in this chapter, in Figure 4.4. The 4G transition toward control and user plane separation (CUPS) splits the SGW into SGW-CP and SGW-UP, for control plane and user plane, respectively. Equally, the PGW will be broken into PGW-CP and PGW-UP, for control and user planes [9].

CUPS concept refers to the control and user plane separation of EPC nodes and provides the architecture enhancements for the separation of functionality in the Evolved Packet Core's SGW, PGW, and TDF. CUPS is defined as of 3GPP Release 14 and paves the way for gradual 5G adaptation via 4G EPC as the concept enables flexible network deployment and operation via distributed or centralized deployment and independent scaling between control plane and UPFs [9].

The benefit of CUPS is the reduced latency on application service. This can be done by relying on user plane nodes closer to the RAN without affecting the number of control plane nodes. CUPS concept also supports increasing data traffic as the service utilization increases as the user plane nodes can be added into the MNO infrastructure without impacting the number of SGW-C, PGW-C, and TDF-C elements of the mobile network.

Other benefits include the possibility to add and scale the EPC node CP and UP resources independently, which means that the evolution of the CP and UP functions can be done independently. CUPS also enables SDN for optimized user plane data delivery.

As depicted in Figure 4.27, CUPS concept has three interfaces, which are *Sxa*, *Sxb*, and *Sxc*. These are located between the CP and UP functions of the SGW, PGW and TDF, respectively.

3GPP has adopted a 3GPP native protocol for the *Sxa*, *Sxb*, and *Sxc* interfaces with TLV-encoded messages over UDP/IP, referred to as packet forwarding control plane (PFCP) protocol, which is depicted in Figure 4.28.

Among various tasks defined by 3GPP, the PFCP includes a heartbeat procedure to check that a PFCP peer is alive, load control, and overload control procedures for balancing the load of different UP functions and reducing signaling to UP function in case there is an overload situation. *Sx* session-related procedures include the *Sx* session establishment, as well as modification and deletion procedures, *Sx* session reporting for traffic usage.

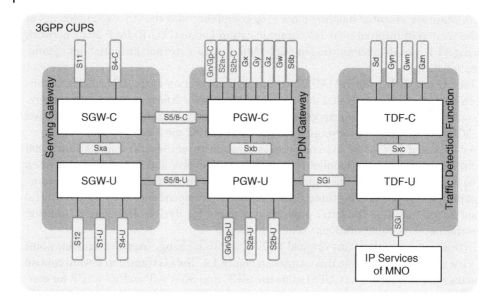

Figure 4.27 Control and user plane separation (CUPS) as defined by 3GPP [9]. CUPS concept supports the increased traffic by allowing the deployment of additional user plane nodes without need to change the amount of SGW-C, PGW-C, and TDF-C elements. The MNO IP services refer to, e.g. IMS and primary synchronization signal (PSS).

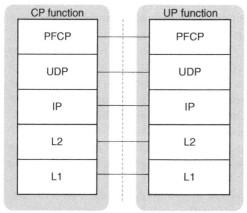

Sx reference point

Figure 4.28 The principle of PFCP protocol.

Also, new DNS procedures are added for the selection of UP function, which is done by CP function based on DNS or local configuration and capabilities of the UP function and the overload control information.

References

1 T-Mobile, 5G Architecture Options - Full Set (RP-161266 of RAN/SA meeting), 3GPP, Busan, 2016.

2 3GPP, 3GPP TSG-SA WG3 Meeting #91, CR S3-181517, 3GPP S3, 2018.

3 3GPP, TS 38.401, 3GPP, 2018.

4 3GPP, TS 23.501. System Architecture for the 5G System, Release 15, V. 15.1.0, 3GPP, 2018.

5 Wilke, J., 5G Network Architecture and FMC Joe Wilke, ITU, 2017.

6 5G Americas, Network Slicing for 5G and Beyond, 5G Americas, 2016.

7 3GPP, System architecture milestone of 5G Phase 1 is achieved, 3GPP, 21 December 2017. http://www.3gpp.org/NEWS-EVENTS/3GPP-NEWS/1930-SYS_ARCHITECTURE. [Accessed 29 July 2018].

8 IETF, Enabling ICN in 3GPP's 5G NextGen Core Architecture, IETF, 29 October 2017. https://tools.ietf.org/id/draft-ravi-icnrg-5gc-icn-00.html. Accessed 29 July 2018.

9 3GPP, "Control and User Plane Separation of EPC nodes (CUPS)," 3GPP, 03 July 2017. http://www.3gpp.org/cups. [Accessed 29 July 2018].

5

Radio Network

5.1 Overview

This chapter presents the new 5G radio interface and the principles in the functioning of the 5G radio network. The introduction gives a description of some of the most relevant theories and technologies related to 5G. Then, an overview of the 5G spectrum is given, including discussion of the feasibility of wider bandwidth utilization in theory and practice, planned 5G radio frequency (RF) bandwidths and band allocations, as well as the latest news and expectations of the World Radio Conference (WRC) of the radio section of International Telecommunication Union (ITU-R) as the forum establishes the foundation of the global utilization and rules of the 5G radio frequencies.

Next, the 5G Radio Access Technologies (RATs) are discussed with an explanation of the principles of established and further expected technologies, advanced multiple-in, multiple-out (MIMO), and carrier aggregation (CA). This section includes discussion of the ITU-R RAT candidates, proposed 3GPP design, and how it addresses the ITU-R Recommendations and use cases for 5G.

This chapter also includes introduction to 5G user devices, and new aspects of 5G terminals. Then, the HW of the radio network equipment is described including the most relevant references for further information. One of the rapidly evolving areas in 5G is the antenna system, so it is explained according to the latest knowledge on evolved adaptive antenna systems and other novelty antenna solutions that are feasible candidates to the 5G networks.

As a last part of this chapter, radio resource utilization is discussed, including eco-friendly equipment evolution, low power, dynamic radio resource utilization, evolution of the OFDM from the 4G era to optimally support the more demanding 5G requirements. Also, the Citizens Broadband Radio Service (CBRS) is summarized as it is expected to open new businesses by widening the operator landscape along with the new non-licensed frequency in the United States.

Amongst multiple specifications, the key documents for this chapter are based on the 38-series of the 3GPP technical specifications.

5G Explained: Security and Deployment of Advanced Mobile Communications, First Edition.
Jyrki T. J. Penttinen.
© 2019 John Wiley & Sons Ltd. Published 2019 by John Wiley & Sons Ltd.

5.2   5G Performance

5.2.1   General

As the 5G is the most versatile set of frequencies ever seen in previous mobile communication generations, also the respective radio network planning will experience major changes. The radio interface of the 5G will experience a major facelift compared to the previous mobile communication systems. The main reason for such a change is the new, completely renovated capabilities of the 5G wireless access that need to be capable of serving use cases that are well beyond previous generations.

The key capabilities include much higher data rates, and much lower latency values. Also, the ability to keep serving the customers with ultra-high reliability, and to support higher energy efficiency and extreme device density, will mean that the radio technology need to be upgraded compared to previous versions, including that of the most advanced Long-Term Evolution (LTE) phase.

The development of the new radio (NR) interface for the 5G will be happening via the further evolution of LTE, combined with completely new radio-access technologies. This native, 5G-specific radio interface is referred to as NR, and along with this evolution, some of the key solutions include the extension of the supported frequencies to higher frequency bands, well beyond 6 GHz region.

There also will be deeper integration of the access and backhaul subsystems. Furthermore, 5G is specifically considering the fluent support of the device-to-device communication, and it thus supports a huge amount of simultaneously communication devices, such as vehicles, machines, intelligent sensors, etc. Yet another key evolution topic of the 5G is the dynamic characteristics of the communications, including flexile duplex and spectrum allocation, relying increasingly on the multi-antenna transmission. It is worth mentioning also the new, clearer separation of the user and control plains [1].

The 5G architecture optimizes the resources and coverage area as it minimizes transmission load, which is not directly related to the user data. The nonuser data include signaling for synchronization, channel estimation, network acquisition, and broadcast of system and control information [1].

The simplified design is essential to serve dense utilization centers containing big amount of network nodes and varying traffic conditions. In addition, the 5G design for the transmission optimizes any deployments. As the network nodes may enter fast, low-energy state upon the load, e.g. when no user-data transmission takes place; it contributes positively to the energy efficiency. An additional benefit of the 5G network design is the positive impact on the high data rate as the interference from others than user data transmissions can be minimized.

5.2.2   Radio Performance Figures

5G, as defined by 3GPP, aims to comply with the strict performance requirements of the International Telecommunication Union's (ITU's) IMT-2020.

Among other advances, also the frame structure of the 5G will be evolving. Nevertheless, the 5G frame structure remains comparable with the 4G. Table 5.1 summarizes the key statements relevant to the frame structure of the radio interface, including a comparison with the previous LTE radio networks.

Table 5.1 The key definitions for the 4G and 5G frame structure and carriers.

Frame structure	4G LTE	5G NR
Radio frame duration	10 ms	10 ms
Subframe duration	1 ms	1 ms
Slot duration	0.5 ms	0.5 ms
Slot format	Predefined	Configurable in a dynamic and semi-statistical way

Table 5.2 The key characteristics comparation of the 4G and 5G radio interface.

Characteristic	4G LTE	5G NR
Channel coding for data	Turbo	LDPC
Channel coding for control	TBCC	Polar
Modulation scheme for uplink	Single-carrier frequency-division multiplexing (SC-FDMA)	DFT-S-OFDM; OFDM (optional)
Modulation scheme for downlink	OFDM	OFDM
Bandwidth (MHz)	1.4, 3, 5, 10, 15, 20	5, ..., 100 (sub-6 GHz); 50, ..., 400 (above 6 GHz)
Subcarrier spacing (kHz)	15 (unicast, Multimedia Broadcast Multicast Service [MBMS]); 7.5/1.25 (MBMS dedicated carrier)	30, 60, 120; 240 (not for data)
Max. carrier aggregation (CC)	32	16
Max. MIMO antenna ports	8 (SU-MIMO); 2 (MU-MIMO)	8 (SU-MIMO); 16 (MU-MIMO)
HARQ transmission/retransmission	TB	TB, Code block group

There are more notable changes in many other items related to the radio interface. Table 5.2 summarizes other relevant key statements of the 4G and 5G radio interfaces.

5.3 5G Spectrum

5.3.1 Overall Advances of 5G Frequencies

The elemental enhancement of 5G is the ability to handle much faster data rates and to provide higher capacity for simultaneously communicating consumers and machines. To cope with these demands and requirements, 5G networks will provide radio equipment with extended support of bands and bandwidths.

Among other advances in the RF, there will be new frequency bands below and above 6 GHz band. The decision of the globally agreed frequencies will be decided in the ITU-R, while the country-specific deployments depend on each areas' regulator bodies. The current discussion includes many variants up to about 100 GHz bands. In practice, the most favorable frequency strategies support as much and big chunks of contiguous bands as possible. The final decision of each region and country is based on the complete picture of the entities requiring and needing the frequencies.

As there is typically more demand than supply for the frequencies, there might be need for optimized utilization of the frequencies. One of the ways to utilize the bands as efficiently as possible is the so-called white space approach. It refers to shared bands that can be utilized by different stakeholders upon the need. In addition to the traditional modes for business models (mobile network operators (MNOs) purchasing right for licensed frequency utilization) and these novel ideas for more optimal performance via the capacity sharing, there are also potential options for further, new-spectrum chunks such as satellite communication and radio location. Some examples of these sharing modes are seen via the Licensed Shared Access (LSA) that is currently under planning in Europe at 2.3 GHz band, as well as the Citizens Broadband Radio Service (CBRS) in the United States that would rely on the 3.5 GHz band.[2]

As the need for the 5G frequency bands increase along with the expected, much higher utilization of the 5G services as ever before within the previous mobile generations – partially due to the huge increase of the M2M type of communications – the propagation characteristics of the radio waves are the bottleneck. Thus, for the largest coverage areas per radio cell, the frequencies need to be low while the highest capacity coverage areas need to rely on the higher frequencies.

It can be summoned up that the most relevant radio frequencies for the 5G range from about 1 up to 30 GHz, but the technically functional, very near-range cells may be relying on the solutions up to about 100 GHz. The highest frequencies provide much-needed capacity for the limited locations and the lowest frequencies ensure the basic functioning of the 5G services within widest areas.

5.3.2 ITU-R WRC-19 Expectations

The next decisions for the 5G frequency strategies are being prepared for the ITU-R WRC-19 event, which is the last occasion prior to the first commercial 5G deployments that will comply with the ITU IMT-2020 requirements. Meanwhile, the previous ITU-R WRC-15 has identified a set of frequencies to be studied for the feasibility for the 5G. The identified frequencies include 24.25–27.50, 37.00–40.50, 42.50–43.50, 45.50–47.00, 47.20–50.20, 50.40–52.60, 66.00–76.00, and 81.00–86.00 GHz. These frequencies are being studied for the use of the mobile service on a primary basis. There are also frequency bands under study requiring possibly additional allocations to the mobile service on a primary basis, and these bands include 31.80–33.40, 40.50–42.50, and 47.00–47.20 GHz.

In practice, the US Federal Communications Commission (FCC) has been considering bands above 24 GHz for 5G. Other regulators are also investigating the options for the preferred bands above 30 GHz for the mobile industry.

5.3.3 5G Bands

The frequency bands for the LTE are defined in the 3GPP TS 36.104, and the 5G bands can be found in the 3GPP TS 38.104. As can be seen in Table 5.3, the number of the

Table 5.3 The frequency bands and frequency ranges for the LTE as interpreted from the 3GPP TS 36.104.

CH	$f_{UL, low}$ MHz	$f_{UL, high}$ MHz	$f_{DL, low}$ MHz	$f_{DL, high}$ MHz	Mode
1	1920.0	1980.0	2110.0	2170.0	FDD
2	1850.0	1910.0	1930.0	1990.0	FDD
3	1710.0	1785.0	1805.0	1880.0	FDD
4	1710.0	1755.0	2110.0	2155.0	FDD
5	824.0	849.0	869.0	894.0	FDD
6	830.0	840.0	875.0	885.0	FDD
7	2500.0	2570.0	2620.0	2690.0	FDD
8	880.0	915.0	925.0	960.0	FDD
9	1749.9	1784.9	1844.9	1879.9	FDD
10	1710.0	1770.0	2110.0	2170.0	FDD
11	1427.9	1447.9	1475.9	1495.9	FDD
12	699.0	716.0	729.0	746.0	FDD
13	777.0	787.0	746.0	756.0	FDD
14	788.0	798.0	758.0	768.0	FDD
15	N/A	N/A	N/A	N/A	FDD
16	N/A	N/A	N/A	N/A	FDD
17	704.0	716.0	734.0	746.0	FDD
18	815.0	830.0	860.0	875.0	FDD
19	830.0	845.0	875.0	890.0	FDD
20	832.0	862.0	791.0	821.0	FDD
21	1447.9	1462.9	1495.9	1510.9	FDD
22	3410.0	3490.0	3510.0	3590.0	FDD
23	2000.0	2020.0	2180.0	2200.0	FDD
24	1626.5	1660.5	1525.0	1559.0	FDD
25	1850.0	1915.0	1930.0	1995.0	FDD
26	814.0	849.0	859.0	894.0	FDD
27	807.0	824.0	852.0	869.0	FDD
28	703.0	748.0	758.0	803.0	FDD
29	N/A	N/A	717.0	728.0	FDD
30	2305.0	2315.0	2350.0	2360.0	FDD
31	452.5	457.5	462.5	467.5	FDD
32	N/A	N/A	1452.0	1496.0	FDD
33	1900.0	1920.0	1900.0	1920.0	TDD
34	2010.0	2025.0	2010.0	2025.0	TDD
35	1850.0	1910.0	1850.0	1910.0	TDD
36	1930.0	1990.0	1930.0	1990.0	TDD
37	1910.0	1930.0	1910.0	1930.0	TDD
38	2570.0	2620.0	2570.0	2620.0	TDD
39	1880.0	1920.0	1880.0	1920.0	TDD

(Continued)

Table 5.3 (Continued)

CH	$f_{UL, low}$ MHz	$f_{UL, high}$ MHz	$f_{DL, low}$ MHz	$f_{DL, high}$ MHz	Mode
40	2300.0	2400.0	2300.0	2400.0	TDD
41	2496.0	2690.0	2496.0	2690.0	TDD
42	3400.0	3600.0	3400.0	3600.0	TDD
43	3600.0	3800.0	3600.0	3800.0	TDD
44	703.0	803.0	703.0	803.0	TDD
45	1447.0	1467.0	1447.0	1467.0	TDD
46	5150.0	5925.0	5150.0	5925.0	TDD
47	5855.0	5925.0	5855.0	5925.0	TDD
48	3550.0	3700.0	3550.0	3700.0	TDD
49	3550.0	3700.0	3550.0	3700.0	TDD
50	1432.0	1517.0	1432.0	1517.0	TDD
51	1427.0	1432.0	1427.0	1432.0	TDD
52	3300.0	3400.0	3300.0	3400.0	TDD
65	1920.0	2010.0	2110.0	2200.0	FDD
66	1710.0	1780.0	2110.0	2200.0	FDD
67	N/A	N/A	738.0	758.0	FDD
68	698.0	728.0	753.0	783.0	FDD
69	N/A	N/A	2570.0	2620.0	FDD
70	1695.0	1710.0	1995.0	2020.0	FDD
71	663.0	698.0	617.0	652.0	FDD
72	451.0	456.0	461.0	466.0	FDD
73	450.0	455.0	460.0	465.0	FDD
74	1427.0	1470.0	1475.0	1518.0	FDD
75	N/A	N/A	1432.0	1517.0	FDD
76	N/A	N/A	1427.0	1432.0	FDD
85	698.0	716.0	728.0	746.0	FDD

LTE bands has increased steadily along with the new releases of 3GPP technical specifications. The presented list is based on the specification version 15.2.0, dated March 2018.

The bandwidth of the LTE can be 1.4, 5, 10, 15, or 20 MHz, depending on the band number. The carrier aggregation provides further means to combine these bands to achieve wider total bandwidth per single user.

The 5G NR frequency bands are defined in the 3GPP 38.104. Table 5.4 summarizes the 5G NR radio frequencies and bands interpreted from the above-mentioned source, version 15.1.0, which is dated March 2018. As can be seen in the table, many bands are shared with the LTE bands (5G bands n1–n76 and 4G LTE bands 1–76), whereas the rest of the 5G NR bands are new (n77–n84 and n257, n258 and n260).

As has been the case with the LTE development, it can be expected that there will be multitude of new 5G operating bands and channel bandwidths as new 3GPP releases are available. Remarkably, the ITU-R WRC-19 will be an important milestone to decide the

Table 5.4 The NR bands and frequency ranges as interpreted from the 3GPP TS 38.104.

CH	$f_{UL, low}$ MHz	$f_{UL, high}$ MHz	$f_{DL, low}$ MHz	$f_{DL, high}$ MHz	Mode
n1	1920.0	1980.0	2110.0	2170.0	FDD
n2	1850.0	1910.0	1930.0	1990.0	FDD
n3	1710.0	1785.0	1805.0	1880.0	FDD
n5	824.0	849.0	869.0	894.0	FDD
n7	2500.0	2570.0	2620.0	2690.0	FDD
n8	880.0	915.0	925.0	960.0	FDD
n20	832.0	862.0	791.0	821.0	FDD
n28	703.0	748.0	758.0	803.0	FDD
n38	2570.0	2620.0	2570.0	2620.0	TDD
n41	2496.0	2690.0	2496.0	2690.0	TDD
n50	1432.0	1517.0	1432.0	1517.0	TDD
n51	1427.0	1432.0	1427.0	1432.0	TDD
n66	1710.0	1780.0	2110.0	2200.0	FDD
n70	1695.0	1710.0	1995.0	2020.0	FDD
n71	663.0	698.0	617.0	652.0	FDD
n74	1427.0	1470.0	1475.0	1518.0	FDD
n75	N/A	N/A	1432.0	1517.0	SDL
n76	N/A	N/A	1427.0	1432.0	SDL
n77	3300.0	4200.0	3300.0	4200.0	TDD
n78	3300.0	3800.0	3300.0	3800.0	TDD
n79	4400.0	5000.0	4400.0	5000.0	TDD
n80	1710.0	1785.0	N/A	N/A	SUL
n81	880.0	915.0	N/A	N/A	SUL
n82	832.0	862.0	N/A	N/A	SUL
n83	703.0	748.0	N/A	N/A	SUL
n84	1920.0	1980.0	N/A	N/A	SUL
n257	26 500.0	29 500.0	26 500.0	29 500.0	TDD
n258	24 250.0	27 500.0	24 250.0	27 500.0	TDD
n260	37 000.0	40 000.0	37 000.0	40 000.0	TDD

global strategy for the utilization of the bands above 6 GHz spectrum while this present list includes the sub-6 GHz bands identified up to date.

3GPP refers this division into sub-6 GHz bands and the bands above 6 GHz as Frequency Range 1 and Frequency Range 2, respectively. More specifically, FR1 covers the frequencies in 450 MHz–6 GHz range while FR2 refers to the frequencies within 24.250–52.600 GHz; as an example, the bands n257, n258, and n260 of Table 5.4 belong to FR2; the rest being in FR1.

The 5G NR is able to support different user equipment (UE) channel bandwidths in a flexible way while it operates within the base station's (BS's) channel bandwidth. As the 3GPP 38.104 states, the base station can transmit to and/or receive from one or more

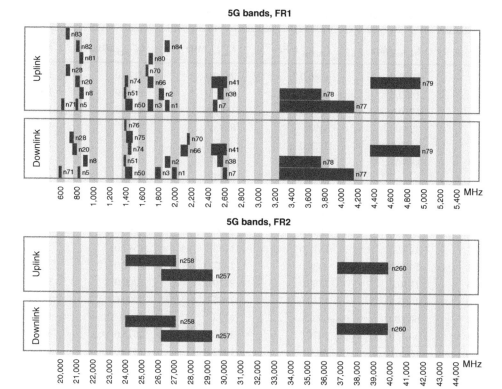

Figure 5.1 The 5G NR bands as defined in 3GPP Release 15.

UE bandwidth parts that are smaller than or equal to the number of carrier resource blocks on the RF carrier, in any part of the carrier resource blocks.

The TS 38.104 specifies multiple transmission bandwidth configurations N_{RB} per base station channel bandwidth and respective subcarrier spacing for FR1 and FR2. The TR1 transmission bandwidth configurations can have bandwidth values of 5, 10, 15, 20, 25, 30, 40, 50, 60, 70, 80, 90, and 100 MHz while the subcarriers can be varied between the values of 15, 30, and 60 kHz. For the FR2 mode, the transmission bandwidth configuration can have bandwidth values of 50, 100, 200, and 400 MHz while the subcarriers can be either 60 or 120 kHz.

Figure 5.1 depicts the 5G bands in graphical format. The more specific requirements for the RF channel utilization, including the guard bands, tolerance values for interfering bands, etc., are found in the 3GPP TS 36.104 and TS 38.104 for 4G LTE and 5G NR, respectively.

5.4 5G Radio Access Technologies

5.4.1 Key Specifications

3GPP defines the 5G radio technology in the 38-series, under the name NR, whereas the core network (CN) is referred to as NGC (next-generation core). Table 5.5 lists some of

Table 5.5 Some of the key technical specifications of 3GPP New Radio (NR) interface.

TS	Title
38.101	User equipment (UE) radio transmission and reception
38.104	Base station (BS) radio transmission and reception
38.201	Physical layer; general description
38.211	Physical channels and modulation
38.300	NR overall description (Stage-2)
38.305	NG Radio Access Network (NG-RAN); Stage 2 functional specification of user equipment (UE) positioning in NG-RAN
38.306	User equipment (UE) radio access capabilities
38.321	Medium Access Control (MAC) protocol specification
38.322	Radio Link Control (RLC) protocol specification
38.323	Packet Data Convergence Protocol (PDCP) specification
38.331	Radio Resource Control (RRC); protocol specification
38.401	NG-RAN; architecture description
38.410	NG-RAN; NG general aspects and principles
38.801	Study on new radio access technology; radio access architecture and interfaces

the fundamental 5G radio technical specifications referenced in this book. The complete list of the 38-series can be found online in [3].

5.4.2 Frequency Bands

For the additional frequency bands and bandwidth variants designed especially for 5G, some of the respective key 3GPP Technical Reports are summarized in Table 5.6.

5.4.3 5G Channel Modeling

As the current radio frequencies are getting increasingly congested while the data rates and capacity demand increase, 5G requires new bands. The mobile communication

Table 5.6 The key technical reports detailing 5G-specific radio frequency bands.

TR	Title
38.812	Study on non-orthogonal multiple access (NOMA) for NR
38.813	New frequency range for NR (3.3–4.2 GHz)
38.814	New frequency range for NR (4.4–4.99 GHz)
38.815	New frequency range for NR (24.25–29.5 GHz)
38.817-01	General aspects for UE RF for NR
38.817-02	General aspects for BS RF for NR
38.900	Study on channel model for frequency spectrum above 6 GHz
38.901	Study on channel model for frequencies from 0.5 to 100 GHz

industry has identified the potential for the range of 6–100 GHz for future 5G networks as it provides means for high capacity and densification of networks. Due to the short communications range, this range is especially suitable for the small-cell deployment scenario.

As an example, the FCC has published rules that provide flexible wireless broadband. FCC allows 3.85 GHz of licensed flexible use at 28–40 GHz bands, and an unlicensed band at 3.5 and 64–71 GHz. There also are plans for adapting 24–25, 32, 42, 48, 51, 70, and 80 GHz for 5G.

The new, much higher frequencies up to 100 GHz require renewed radio propagation models, too. One of the related key studies is found in the 3GPP TR 38.900.

There also have been various channel measurements and modeling efforts such as METIS2020, COST2100, European Telecommunications Standards Institute (ETSI) mmWave SIG, MiWEBA, mmMagic, NYU WIRELESS, and Globecom 2015. The 5G channel modeling is typically based on measurements and ray-tracing concept.

5.4.4 Radio Technology Principles

This section outlines the expected key radio technologies for the 5G that may contribute to achieve the high capacity and radio performance to comply with the strict ITU-R IMT-2020 requirements.

5.4.4.1 OFDM in 5G

The 3GPP has chosen CP-OFDM (Cyclic Prefix-Orthogonal Frequency Division Multiplexing) waveform for 5G NR. The NR and previously defined LTE form jointly the 5G radio access, supporting each other especially in the initial phase of the 5G deployment. While LTE continues supporting frequency bands below 6 GHz, the NR will have a variety of bands over a wide spectrum, from sub-1 up to 100 GHz. The benefit of the joint functioning of the LTE and NR is the possibility of achieving capacity gain via aggregation.

Benefits of OFDM-Based Modulation The Orthogonal Frequency Division Multiplexing (OFDM) modulation has been selected as a base for 5G for several reasons. The following summarizes the key justifications [4]:

- OFDM is spectral efficient both in uplink (UL) and downlink (DL) to comply with high data rate requirements. In addition to the radio interface as such, also backhaul benefits from OFDM-based modulation technique, as well as, e.g. dense urban vehicular communication use case where several vehicles are performing asynchronous broadcasting.
- OFDM provides means for fluent utilization of MIMO, which provides high spectral efficiency via both single-user MIMO (SU-MIMO) and multi-user MIMO (MU-MIMO). Combined with adaptive beamforming technology, the OFDM contributes to the compensation of the radio propagation loss on high-frequency bands.
- OFDM can be optimized for better applicability to peak-to-average-power-ratio (PAPR). In the previous generation's LTE and LTE-Advanced systems, the OFDM has been selected into downlink while the uplink relies on single-carrier

frequency-division multiplexing access (SC-FDMA), which is more adequate in optimizing device's power efficiency. The PAPR of OFDM in 5G can be lowered by applying PAPR reduction techniques with minor impact in performance [5].

- The high-speed use cases require robustness in channel time-selectivity. OFDM supports this via adequate adjustment of subcarrier spacing. It can be assumed that the highest frequency bands will be utilized for small cells. Along with 5G, high robustness is required for supporting the fluent vehicle communications in V2X. It can also be expected that the importance of mobile backhaul will increase.
- OFDM system can be made robust to phase noise by a proper choice of subcarrier spacing.
- The baseband complexity of an OFDM receiver is low.
- OFDM is well localized in time domain, which is relevant in the support of latency-critical ultra-reliable low latency communications (URLLC) and dynamic Time Division Duplex (TDD). The respective use cases include, e.g. the backhaul transmission and V2X communications. Nevertheless, OFDM is not performing optimally in frequency domain. Frequency localization may be important in use cases supporting coexistence of different services having separate waveform types in frequency domain on shared carrier, but in general, frequency localization of a waveform may not be of utmost importance on higher frequencies with available bandwidth.
- The OFDM cyclic-prefix (CP) makes it robust to timing synchronization errors.
- OFDM is a flexible waveform, so it supports a variety of use cases and services over wide range of frequencies when subcarrier spacing and cyclic prefix are adjusted adequately.

OFDM Principle After the evaluation of new candidates to 5G waveforms, the OFDM – familiar from the LTE and LTE-Advanced – was selected as its performance has been proven in practice. It can be further optimized to tackle the strict 5G requirements. The additional benefit of this selection is that for those familiar with the OFDM via LTE (or via many other environments such as Wi-Fi radio interface), there are some minor additions to the already adapted principles. This section presents the OFDM as an elemental base for the 5G, too.

The OFDM refers to the technology that divides a wide frequency band into various narrow frequencies, i.e. subcarriers that carry the actual data between the transmitter and receiver. The multiplexing takes care of the simultaneous transmission of the data on each individual subcarrier. In the LTE and LTE-Advanced systems, the subcarriers have a fixed width of 15 kHz, and depending on the need for the data speed, each device is dedicated dynamically with a variable set of these subcarriers. The idea of using the subcarriers is to obtain a radio channel that is roughly constant (flat) over each given sub-band, which minimizes the negative effects of the fading, as the faded band is typically very narrow and the subcarrier can be reallocated during the fading. This principle makes the equalization much simpler, at the receiver compared to the previous techniques utilized in 2G, 3G, and 4G.

The OFDM symbol duration time is $1/\Delta f$ + cyclic prefix (CP). The cyclic prefix is used to maintain orthogonality between the subcarriers even for a time-dispersive radio channel. In the LTE, a single resource element carries the data by using either

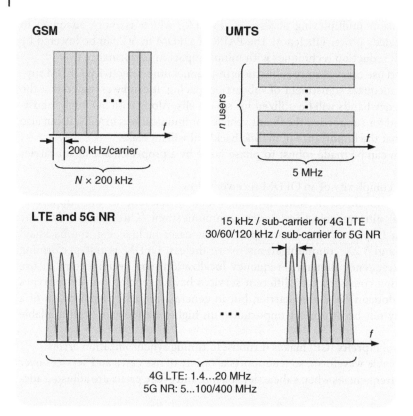

Figure 5.2 The frequency band of OFDM as applied into LTE and 5G NR consists of several subcarriers while the WCDMA of UMTS utilizes one complete carrier for all the code division traffic of a single cell. The LTE and 5G NR allow the additional capacity via carrier aggregation. As a comparison, the GSM system is based on TDMA.

Quadrature Phase Shift Keying (QPSK), 16QAM (Quadrature Amplitude Modulation), or 64QAM modulation.

Figure 5.2 depicts the difference between 3G with a fixed bandwidth of 5 MHz, which is shared between all the admitted users based on code division multiplexing, and the flexible LTE bandwidth, which has finer granularity for defining the frequency band; it is possible to deploy 1.4, 3, 5, 10, 15, or 20 MHz bands in the LTE, and further combine the bands via carrier aggregation concept. As can be seen, one of the benefits of OFDM over Code Division Multiple Access (CDMA) is the dynamic allocation of the subcarriers, which can optimize the frequency band re-farming between LTE and other systems as the utilization of previous systems will decrease. The figure also presents the TDMA (Time Division Multiple Access) principle of the Global System for Mobile Communications (GSM), which, in fact, is a combination of TDMA and Frequency Division Multiple Access (FDMA) formed by eight time slots, or their subdivided resources per each 200 kHz carrier.

For 5G NR, the bandwidth can vary in a wider way, from 5 up to 100 or 400 MHz, depending on the scenario.

OFDM is based on the Frequency Division Multiplexing (FDM). In FDM, different streams of information are mapped onto separate parallel frequency channels. OFDM differs from traditional FDM in terms of the following aspects:

- The same information stream is mapped onto many narrowband subcarriers, increasing the symbol period compared to single carrier schemes.
- The subcarriers are orthogonal to each other to reduce the inter-carrier interference (ICI). Moreover, overlap between subcarriers is allowed to provide high spectral efficiency.
- A guard interval, often called *cyclic prefix*, is added at the beginning of each OFDM symbol to preserve orthogonality between subcarriers and eliminate inter-symbol interference (ISI) and ICI (see Figures 5.3 and 5.4).

In the frequency domain, the overlap between subcarriers can take place as they are orthogonal to each other.

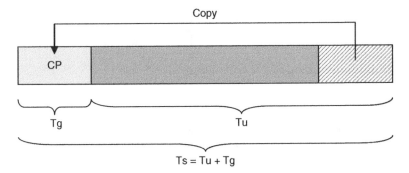

Figure 5.3 The principle of the OFDM cyclic prefix.

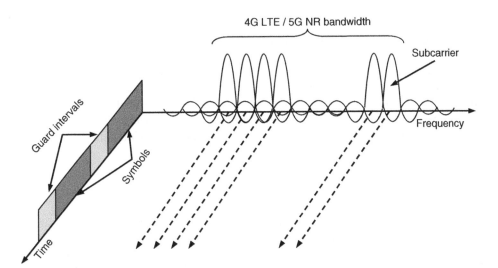

Figure 5.4 Frequency-time interpretation of an OFDM signal.

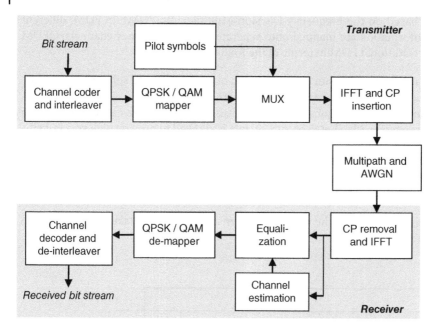

Figure 5.5 SISO OFDM simplified block diagram.

5.4.5 OFDM Transceiver Chain

Figure 5.5 presents a simplified block diagram of a single-input single-output (SISO) OFDM system. On the transmitter side, the modulated (QAM/PSK) symbols are mapped onto N orthogonal subcarriers. This is accomplished by means of an Inverse Discrete Fourier Transform (IDFT) operation. Most commonly, the IDFT is performed with an Inverse Fast Fourier Transform (IFFT) algorithm, which is computationally efficient. Next, the CP is inserted, and a parallel-to-serial conversion is performed prior to the transmission over the air.

At the receiver end, the reversal operations are performed. Once the received signal reaches the receiver, the CP, which is potentially interfered with by previous OFDM symbols, is removed. Then, a fast Fourier transform (FFT) operation brings the data to the frequency domain. This way, channel estimation and equalization are simplified. Note that to be able to carry out the latter operations, known symbols called *pilots* are to be inserted in certain frequency positions/subcarriers at the transmitter side. At the end of the chain, the equalized data symbols are demodulated yielding the received bit stream.

5.4.6 Cyclic Prefix

A guard interval is added in the beginning of each OFDM symbol to minimize negative impact of the multipath channel. If the duration of the guard interval T_g is larger than the maximum delay of the channel τ_{max}, all multipath components will arrive within this guard time and the useful symbol will not be affected avoiding ISI as can be seen in Figure 5.6.

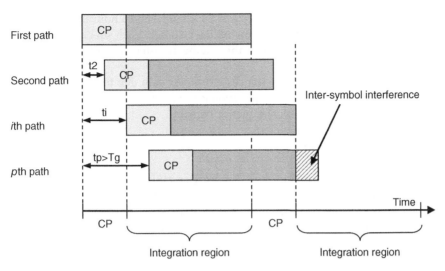

Figure 5.6 Cyclic prefix (CP) avoiding ISI.

One particularization of the guard interval is the so called cyclic prefix. In this case the last N_g samples of the useful OFDM symbol with N samples in total are copied to the beginning of the same symbol. Since the number of cycles of each orthogonality function per OFDM symbol will be maintained as an integer, this strategy also allows to keep the orthogonality properties of the transmitted subcarriers avoiding ICI. Figure 5.6 shows the cyclic prefix concept where

$$T_u = N \times T_0$$

$$T_g = N_g \times T_0$$

$$T_s = (N + N_g) \times T_0$$

The variables refer to the following: T_u is the useful OFDM symbol where data symbols are allocated, T_g is the duration of the cyclic prefix and T_s is the total duration of the OFDM symbol. The insertion of the CP results in Spectral Efficiency Loss (SEL), which is not relevant compared to the benefits that provides in terms of ISI and ICI robustness. The SEL can be interpreted as the loss of throughput that the OFDM transmission system will suffer by the addition of the cyclic prefix, and it can be presented as:

$$SEL = \frac{T_g}{T_g + T_u}$$

The loss of spectral efficiency is directly related to the ratio between the duration of the CP and the total duration of an OFDM symbol.

5.4.7 Channel Estimation and Equalization

In wireless OFDM systems, the received symbols have been corrupted by the multipath channel. To undo these effects, an equalization of the received signal that somehow compensates the variations introduced by the channel must be performed.

If the CP is longer than the maximum delay of the channel and a nonvariant channel over the duration of an OFDM symbol (slow-fading channel), each subcarrier symbol is multiplied by a complex number equal to the channel transfer function coefficient at this subcarrier frequency.

In other words, each subcarrier experiences a complex gain due to the channel. To undo these effects a single complex multiplication is required for each subcarrier yielding low complexity equalization in the frequency domain:

$$y[k] = \frac{z[k]}{h[k]} = d[k] + \frac{w[k]}{h[k]}$$

where $y[k]$ is the equalized symbol in the kth subcarrier, $z[k]$ is the received symbol at the kth subcarrier after FFT and $h[k]$ is the complex channel gain at subcarrier k. $w[k]$ represents the additive white Gaussian noise at subcarrier k.

Note that this equalization has been performed assuming a perfect knowledge about the channel. However, in most of systems that employ equalizers, the channel properties are unknown a priori. Therefore, the equalizer needs a channel estimator that provides the equalization block the required information about the channel characteristics.

To estimate the channel in OFDM systems, *pilot-aided channel estimation* is the most suitable solution for the mobile radio channel. This technique consists in transmitting symbols, often called pilot symbols, known by both the transmitter and the receiver to estimate the channel at the receiver. This approach presents an important trade-off between the number of pilots used to perform the estimation and the transmission efficiency. The more pilots are used, the more accurate the estimation will be, but also the more overhead will be transmitted reducing the data rate.

As an example, Figures 5.7–5.9 depict the mapping of cell-specific reference signals in LTE for different number of antenna ports and with normal CP. The pilot symbols are distributed in frequency- and time-domain and they are orthogonal to each other to provide accurate channel estimation. Figure 5.8 shows the idea of the LTE radio resource block, and following figures show the mapping of the reference signals (Figure 5.10).

In 5G, there also exists 8×8 port configuration.

5.4.8 Modulation

LTE can use QPSK, 16QAM, and 64QAM modulation schemes as shown in Figure 5.11, whereas 5G has option to utilize up to 254QAM. The channel estimation of OFDM is usually done with the aid of pilot symbols. The channel type for each individual OFDM subcarrier corresponds to the flat fading. The pilot-symbol assisted modulation on flat fading channels involves the sparse insertion of known pilot symbols in a stream of data symbols.

The QPSK modulation provides the largest coverage areas but with the lowest capacity per bandwidth. 64-QAM results in a smaller coverage, but it offers more capacity.

5.4.9 Coding

LTE uses Turbo coding or convolutional coding, the former being more modern providing in general about 3 dB gain over the older and less effective, but at the same time more

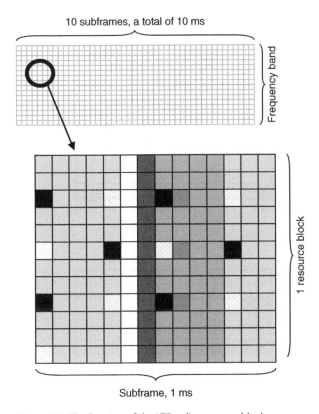

Figure 5.7 The forming of the LTE radio resource block.

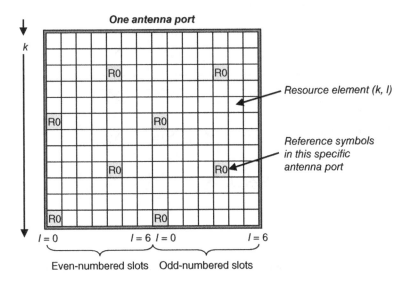

Figure 5.8 Mapping of downlink cell-specific reference signals in LTE with normal CP, i.e. in one antenna port setup of LTE.

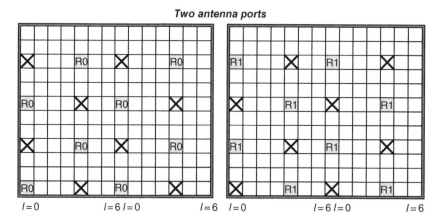

Figure 5.9 Two-port MIMO in LTE. The cross indicates the resource elements that are not used in the respective antenna port.

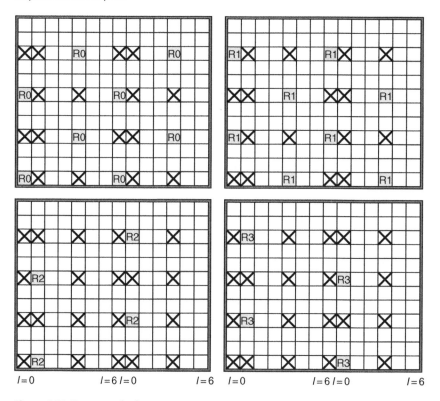

Figure 5.10 Port setup for four antennas.

robust convolutional coding. 5G further optimizes the coding by introducing coding schemes summarized in Table 5.2.

The creation of the OFDM signal is based on the IFFT, which is the practical version of the discrete Fourier transform (DFT) and relatively easy to be deployed as there are

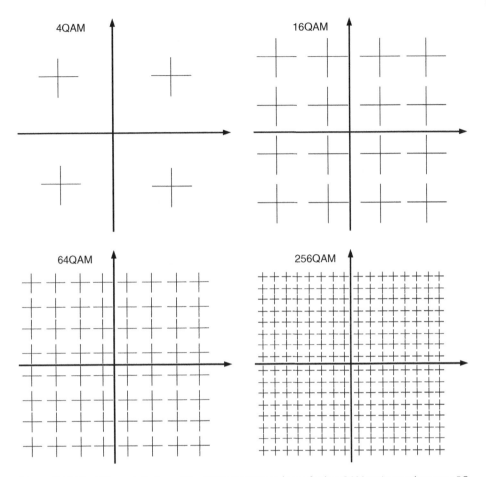

Figure 5.11 The I/Q constellation of the QPSK (4QAM) and set of other QAM variants relevant to 5G.

standard components for the transform calculation. The reception utilizes, on the other side, the FFT for combining the original signal.

5.4.10 Signal Processing Chain

After the coding and modulation of the user data, the OFDM signal is formed by applying serial-to-parallel conversion. This is an essential step to feed the IFFT process. Before bringing the parallel subcarriers of the user data, the subcarrier mapping also takes the needed amount of parallel subcarriers from the other users, i.e. the OFDMA (OFDM Access) is applied. All these streams are fed into the IFFT input in order to do the inversed discrete Fourier transform in a practical way. It is important to note that the process from the serial symbol stream to S/P conversion, sub-carrier mapping process and N-point IFFT process happens in the frequency domain, whereas the process from the IFFT conversion happens in the time domain.

The OFDM symbols are formed by adding the cyclic prefix in the beginning of the symbols in order to protect the signal against multi-path propagated components.

Then, the windowing, digital to analogue conversation, frequency up-conversion, RF processing, and finally the actual radio transmission are performed in the transmitter of the eNodeB. The OFDM transmission is only used in the downlink, so the Long Term Evolution-User Equipment (LTE-UE) does have the OFDM receiver and SC-FDMA transmitter.

5.5   Uplink OFDM of 5G: CP-OFDM and DFT-s-OFDM

As stated in 3GPP TS 38.300, the downlink transmission waveform of the 5G radio interface is conventional OFDM using a cyclic prefix, which is referred to as CP-OFDM. It is the very same as in LTE and LTE-A, as described in the previous section. The uplink transmission waveform of 5G, in turn, is conventional OFDM using a cyclic prefix with a transform precoding function performing DFT spreading that can be disabled or enabled. The latter is referred to as DFT-s-OFDM.

The difference between the 5G and the LTE/LTE-A multiplexing is thus for the uplink; instead of previously utilized SC-TDMA, the 5G is based on OFDM in both downlink and uplink. The DFT spreading that can be applied in 5G uplink optimizes the PAPR performance which was the reason to select the SC-TDMA in the first hand in LTE/LTE-A as the sole conventional OFDM is not optimal for it, especially in the use cases requiring low battery consumption such as low-powered IoT devices.

The requirements of the OFDM in 5G has been defined in 3GPP TS 38.300, Section 5.1. It summarizes the flow of waveform generation for both downlink and uplink (Figure 5.12).

5.6   Downlink

5G supports a closed loop DMRS (Demodulation Reference Signal) -based spatial multiplexing for the Physical Downlink Shared Channel (PDSCH). Furthermore, up to 8 and 12 orthogonal DL DMRS ports are supported for type 1 and type 2 DMRS, respectively. Up to eight orthogonal DL DMRS ports for each UE are supported for SU-MIMO and up to four orthogonal DL DMRS ports for each UE are supported for MU-MIMO.

There is a precoded matrix applied in the transmission of the DMRS and corresponding PDSCH. There is no need for UE to know the precoding matrix to demodulate the transmission.

As defined in 3GPP TS 38.300, the downlink physical-layer processing of transport channels includes the following steps:

- Transport block CRC (cyclic redundancy check) attachment;
- Code block segmentation and code block CRC attachment;
- LDPC (low-density parity check) based channel coding;
- Physical-layer hybrid-ARQ processing and rate matching;
- Bit-interleaving;
- QPSK, 16QAM, 64QAM, and 256QAM modulation schemes;
- Layer mapping and precoding;
- Mapping to assigned resources and antenna ports.

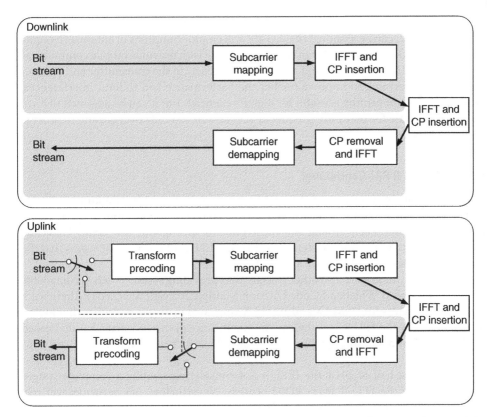

Figure 5.12 The principle of transmitter block for CP-OFDM. The utilization of DFT-spreading in uplink is optional.

To schedule DL transmissions on PDSCH and UL transmissions on Physical Uplink Shared Channel (PUSCH), the UE-specific Physical Downlink Control Channel (PDCCH) is applied. The respective Downlink Control Information (DCI) on PDCCH contains downlink assignments with modulation and coding format, resource allocation, and Downlink Shared Channel (DL-SCH)-related hybrid-ARQ information. It also contains uplink scheduling grants with modulation and coding format, resource allocation, and Uplink Shared Channel (UL-SCH's) hybrid-ARQ information. In the 5G system, the synchronization signal and Physical Broadcast Channel (PBCH) block have primary synchronization signal and secondary synchronization signals (PSS, SSSs).

The UE can rely on band-specific sub-carrier spacing for the SS/PBCH block by default, until network states otherwise.

More detailed description on the DL-SCH physical layer model can be found in the 3GPP TS 38.202, and the PBCH physical layer model is described in 3GPP TS 38.202.

5.6.1 Advanced MIMO

MIMO antenna systems are an integral part of the current 4G deployments. Thus, the evolution of the MIMO can be assumed to continue also in 5G era.

The MIMO antenna systems provide with highly directive antenna pattern forming in a dynamic way. It can be assumed that 5G NR will employ hundreds of antenna elements to increase the performance; this may not be possible with previous mobile communications systems. The MIMO concept also is related closely to the transmitter and receiver antenna beamforming to improve further the performance and to limit interferences. Not only the beamforming is useful for high frequencies, but it can be assumed to form an important base for many low-frequency scenarios to extend coverage and to provide higher data rates.

5.6.2 The ITU-R RAT Candidates

Along with the convergence of the mobile communications, the amount of diverse systems per mobile communications generation has fallen. The systems based on 3GPP specifications have been among the most popular throughout the GSM, Universal Mobile Telecommunications System (UMTS), and LTE evolution, and it seems that the same scenario is continuing as we approach 5G.

In 3G and 4G, there were also other stakeholders providing systems such as cdma2000, an Asian TD-based Wideband Code Division Multiple Access (WCDMA). There were also IEEE standards for WiMAX and WiMAX2 included in the 3G and 4G technologies, as referenced by the ITU. Nevertheless, the significance of these systems has decreased while the LTE and its evolution by LTE-Advanced has increased in relevance and popularity.

Up to the first half of 2018, the IEEE is not considering sending a complete system specification to the ITU-R IMT-20202 process. Nevertheless, 5G networks will rely largely on various protocols defined by IEEE.

5.7 New Radio (NR) Interface of 3GPP

The 3GPP is one of the most concrete standardization bodies driving for the 5G and developing the current 4G technical specifications further for the ITU-R IMT-2020 compliance. This section thus discusses the current understanding of the proposed 3GPP design and how it addresses the ITU-R requirements and use cases for 5G.

The key specification for the NR is the 3GPP 38.300 [6]. Among other aspects, it defines the following key functionalities and principles:

- Protocol architecture and functional split
- Interfaces
- Radio protocol architecture
- Channels and procedures in uplink and downlink
- Medium Access Control (MAC), Radio Link Control (RLC), Packet Data Convergence Protocol (PDCP) and Radio Resource Control (RRC) layers
- Mobility and states
- Scheduling
- UE key functionalities and capabilities
- Quality of service (QoS)
- Security
- Self-configuration and self-optimization

The following sections discuss these aspects apart from the security aspects, which is dealt with more details in a separate chapter later in this book.

5.7.1   Radio Network Architecture and Interfaces

The 3GPP Technical Specification 38.300 defines the 5G radio interface and functionality. The radio network of the 5G has its evolution path. The intermediate step is called non-standalone (NSA) with the base station element called **ng-eNB**. As the 3GPP specifications state, it is a node providing E-UTRA (Evolved UTRA) user plane and control plane protocol terminations toward the UE. It is connected via the Next Generation (NG) interface to the 5G Core Network (5GC). This is thus a 4G eNodeB, which can communicate with the 5G infrastructure.

In the fully evolved standalone (SA) phase, the native 5G radio base station is referred to as **gNB**. It is a node that provides the NR user plane and control plane protocol terminations toward the UE. It is connected via the NG interface to the 5G core network.

The general term for the 5G radio base station is NG-Radio Access Network (NG-RAN) node. It can be either a gNB (which refers effectively to the native 5G base station) or an ng-eNB (which is the intermediate radio base station based on the 4G era). These gNB and ng-eNB elements are interconnected via the *Xn* interface within the radio network. Both elements are connected furthermore via the NG interfaces to the 5G core network. This interface is divided into two parts: user and control interfaces. The control interface is referred to as NG-C, and it connects the radio base stations to the AMF (Access and Mobility Management Function). The user interface is referred to as *NG-U* and it connects the radio base stations to the UPF (user plane function). These interfaces are described in the 3GPP TS 23.501, whereas the functional interface for the control and user plane split as well as the 5G architecture are explained in the 3GPP TS 38.401.

In 5G, the further optimization of the radio resources takes place by separating user and control communications. It refers to the decoupling user data and control planes. This also provides the means to separate the scaling of user plane capacity and control functionality. One example of this is a situation in which user data might be delivered via a dense access node layer while the system information-related messages are delivered via overlaying macro layer.

This separation applies also over multiple frequency bands and RATs. In 5G, this makes it possible to deliver the user data via a dense, high-capacity 5G layer on higher frequency whereas the overlaid LTE layer provides the reliable signaling for call control.

5.7.2   5G Network Elements

5.7.2.1   gNB and ng-NB
Figure 5.13 depicts the NG-RAN architecture of 5G.

The gNB (standalone 5G NodeB) and ng-eNB (non-standalone 4G NodeB) host the following functions:

- Radio Resource Management (RRM)
- IP header management
- AMF management
- Routing functionalities

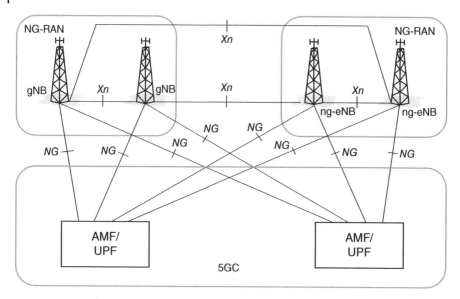

Figure 5.13 The 5G radio network architecture.

- Connection setup and release
- Scheduling functionality
- Measurements
- Packet marking
- Session management
- Network slicing
- QoS functions
- Support of UEs in RRC_INACTIVE state
- Distribution function for NAS messages
- Radio access network sharing
- Dual connectivity
- Tight interworking between NR and E-UTRA

The RRM of the gNB and ng-eNB includes radio bearer and radio admission control, connection and mobility control, and scheduling, which refers to the dynamic resource allocation to a set of UEs in both uplink and downlink. Furthermore, the 5G gNB's IP header management provides data compression, encryption, and integrity protection.

The gNB connects to AMF. The related AMF management from the gNB side refers to the selection of an AMF upon UE attachment in those scenarios when AMF cannot determine the routing information from the UE messaging.

Routing functionality of the 5G NB refers to the routing of user plane data toward a set of UPF elements, and routing of control plane information to AMF.

The connection setup and release performed by the 5G gNB relate to the procedures for initiating and terminating data sessions.

The scheduling functionality of the 5G gNB refers to the ability to schedule and transmit paging messages originated from the AMF, and schedule and transmit system broadcast information originated from the AMF or operations and management system.

The 5G gNB can perform radio interface measurements and deliver respective reports, which assist in the configuration for mobility and scheduling.

The packet marking of the 5G NB refers to the ability to mark the transport level packets in the uplink.

The session management refers to the procedures and functionalities applied during the active data connection.

The network slicing refers to the forming of highly dynamic virtual "sub-networks" with varying capabilities and optimized resource utilization according to the service-based architecture of 5G, taking advantage of the network functions virtualization. In other words, network slicing only consumes those resources from the virtualized environment needed for the specific moment.

The QoS functions refer to the QoS flow management and mapping to data radio bearers.

5.7.2.2 AMF

The AMF refers to as Access and Mobility Management Function. It contains the following key functions as referred in 3GPP TS 23.501:

- NAS signaling termination
- NAS signaling security
- AS security control
- Inter CN node signaling for mobility between 3GPP access networks
- Idle mode UE reachability (including control and execution of paging retransmission)
- Registration area management
- Support of intra-system and inter-system mobility
- Access authentication
- Access authorization including check of roaming rights
- Mobility management control (subscription and policies)
- Support of network slicing
- Session Management Function (SMF) selection

5.7.2.3 UPF

The UPF refers to user plane function. It can perform the following key functions as described in 3GPP TS 23.501:

- Acts as an anchor point for intra- and inter-RAT mobility
- Is external packet data unit (PDU) session point of interconnect to data network
- Performs packet routing and forwarding, packet inspection, and acts in user plane part of policy rule enforcement
- Forms traffic usage reports
- Is uplink classifier supporting routing traffic flows to a data network
- Is branching point supporting multi-homed PDU session
- Manages QoS handling for user plane
- Performs uplink traffic verification, being SDF (service data flow) to QoS flow mapping
- Makes downlink packet buffering and triggers downlink data notification

Table 5.7 The modulation schemes of 5G.

	BPSK	QPSK	16QAM	64QAM	256QAM
Downlink		✓	✓	✓	✓
Uplink, OFDM, and CP		✓	✓	✓	✓
Uplink, DFT-s-OFDM, and CP	✓	✓	✓	✓	✓

5.7.2.4 SMF

The SMF hosts the following main functions, as described in 3GPP TS 23.501:

- Manages session
- Allocates and manages UE IP addresses
- Selects and controls UP function
- Configures traffic steering at UPF routing traffic to proper destination
- Manages control part of policy enforcement and QoS
- Takes care of downlink data notification

5.7.3 Modulation

In 5G, the radio interface supports a set of modulations. For the proper selection of the modulation scheme, a modulation mapper is applied. It receives binary values (0 or 1) as input and produces complex-valued modulation symbols as output. These output modulation symbols can be $\pi/2$-BPSK (Binary Phase Shift Keying), BPSK, QPSK, 16QAM, 64QAM, and 256QAM. The modulation schemes supported are listed in Table 5.7.

5.7.4 Frame Structure

This section outlines the 5G channels and procedures both in uplink and downlink as defined by 3GPP. This section also summarizes the functioning of the mobility and states as well as the scheduling based on the 3GPP TS 38.211 (NR; physical channels and modulation) [7].

In 5G, downlink and uplink transmissions are organized into frames, which are derived from the OFDM structure. A single frame has a duration of

$$T_f = (\Delta f_{max} N_f / 100) \cdot T_c = 10 \, \text{ms}$$

Each frame has 10 subframes. An individual subframe has a duration of

$$T_{sf} = (\Delta f_{max} N_f / 1000) \cdot T_c = 1 \, \text{ms}$$

The number of consecutive OFDM symbols per subframe varies, depending on the number of the symbol and slot. Furthermore, each frame is divided into two half-frames of five subframes each, with half-frame 0 consisting of subframes 0–4 and half-frame 1 consisting of subframes 5–9. More detailed description of the frame structure and frequency band allocation can be found in 3GPP TS 38.133 and TS 38.213.

5.7.5   Physical Channels

5.7.5.1   Uplink

As defined in [7], an *uplink physical channel* refers to a set of resource elements carrying information that originates from higher layers. 3GPP has defined the following *uplink physical channels* for 5G:

- Physical Uplink Shared Channel (PUSCH)
- Physical Uplink Control Channel (PUCCH)
- Physical Random-Access Channel (PRACH)

The physical layer uses *uplink physical signals* that do not carry information originating from higher layers. These uplink physical signals are the following:

- Demodulation reference signals (DM-RSs)
- Phase-tracking reference signals (PT-RSs)
- Sounding reference signals (SRS)

The UE uses a frame structure and physical resources as dictated by [7] when it is transmitting in uplink. Furthermore, [7] defines a set of antenna ports that are applied in uplink.

5.7.5.2   Downlink

As defined in [7], a *downlink physical channel* refers to a set of resource elements that carry information arriving from higher layers. 5G defines the following downlink physical channels:

- Physical Downlink Shared Channel (PDSCH)
- Physical Broadcast Channel (PBCH)
- Physical Downlink Control Channel (PDCCH)

A *downlink physical signal* refers to a set of resource elements that the physical layer uses, yet these do not carry information originating from higher layers. The downlink physical signals are the following:

- Demodulation reference signals (DM-RS) for PDSCH and PBCH
- Phase-tracking reference signals (PT-RS)
- Channel-state information reference signal (CSI-RS)
- Primary synchronization signal (PSS)
- Secondary synchronization signal (SSS)

5.7.6   General Protocol Architecture

The physical layer channels overview is presented in TS 38.201 (NR; general description) and TS 38.202 (NR; services) provided by the physical layer [8]. These specifications also detail 5G protocol architecture and the functional split of it, and the overall and radio protocol architectures, MAC, RLC, PDCP, and RRC.

The 3GPP specifications describe the NR interface covering the interface between the UE and the network on layer 1, 2, and 3. The TS 38.200 series describes the layer 1 (physical layer) specifications. Layers 2 and 3 are described in the TS 38.300 series.

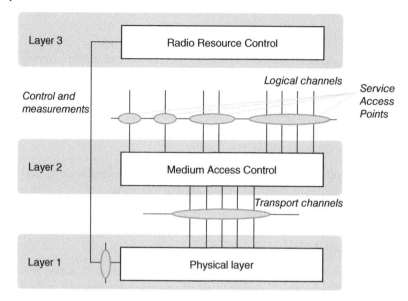

Figure 5.14 The principle of mapping 5G channels.

Figure 5.14 depicts the NR radio interface protocol architecture related to layer 1, which interfaces the MAC layer 2 and the RRC layer 3. The connectivity between layers is done by utilizing service access points (SAPs).

The transport channels between physical radio layer and MAC layer answer the question on how the information is transferred over the radio interface. In the upper layer, between MAC and RRC layers, MAC provides logical channels. They answer the question of what type of information is transferred.

The physical layer offers data transport services to higher layers. The access to these services happens using transport channel through the MAC layer.

5.7.7 Multiple Access

The multiple access scheme for the NR physical layer is based on OFDM with a CP. For uplink, discrete Fourier transform-spread-OFDM (DFT-s-OFDM) with a CP is also supported. To support transmission in paired and unpaired spectrum, both frequency division duplex (FDD) and TDD are enabled.

Layer 1 is based on resource blocks. This provides means for the 5G NR layer 1 to utilize multitude of spectrum allocations. Resource block is based on up to 12 subcarriers.

The duration of radio frame is 10 ms. It contains 10 subframes, and the duration of each is 1 ms. A subframe has one or more adjacent slots, from which each has 14 adjacent symbols.

5.7.8 Channel Coding

The channel coding scheme for transport blocks is quasi-cyclic LDPC codes with two base graphs and eight sets of parity check matrices for each base graph, respectively. One base graph is used for code blocks larger than certain sizes or with initial transmission code rate higher than thresholds; otherwise, the other base graph is used. Before the LDPC coding, for large transport blocks, the transport block is segmented into multiple code blocks with equal size. The channel coding scheme for PBCH and control information is polar coding based on nested sequences. Puncturing, shortening, and repetition are used for rate matching. Further details of channel coding schemes are specified in [5].

5.7.9   Physical Layer Procedures

The physical layer procedures include cell search, power control, uplink synchronization, uplink timing control, random access, and Hybrid Automatic Repeat and Request (HARQ) procedures, antenna beam management, and CSI-related procedures. NR provides support for interference coordination via physical layer resource control in frequency, time, and power domains.

5.7.10   Physical Layer Measurements

Radio characteristics are measured by the UE and the network and reported to higher layers. These include, e.g. measurements for intra- and inter-frequency handover, inter RAT handover, timing measurements, and measurements for RRM. Measurements for inter-RAT handover are defined in support of handover to E-UTRA.

5.7.11   Quality of Service

The 5G aspects on the QoS are defined in 3GPP TS 38.300.

5.8   User Devices

Technical specification 3GPP TS 38.306 V15.1.0 defines the 5G UE radio access capabilities as of Release 15, including new aspects of 5G terminals, local connectivity, evolved battery, display, sensor, memory and processor technologies. Other relevant specifications for the UE's point of view include:

- 3GPP TS 38.300; UE key functionalities and capabilities
- 3GPP TS 38.101-1; NR UE radio transmission and reception, Part 1, Range 1 Standalone
- 3GPP TS 38.101-2; NR UE radio transmission and reception, Part 2, Range 2 Standalone
- 3GPP TS 38.101-3; NR UE radio transmission and reception, Part 3, Range 1 and Range 2 Interworking Operation with Other Radios
- 3GPP TS 38.101-4; NR UE radio transmission and reception, Part 4, Performance Requirements

It should be noted that, as indicated in 3GPP TS 38.300, the UE capabilities in NR do not rely on UE categories. In fact, UE categories indicating peak data rates are only meant for marketing purposes and there is no respective signaling to the 5G network. In practice, the network determines the UL and DL data rate of UE based on the supported band combinations and baseband capabilities, which refer to modulation scheme, MIMO layers, and other characteristics on the RF path.

5.9 Other Aspects

The 3GPP TS 38.300 summarizes some additional aspects respective to 5G, such as self-configuration and self-optimization. It also discusses key security aspects of the 5G, which will be discussed in more detail in Chapter 8.

Along with the 5G deployments, it is expected that more advanced antenna technologies will be taken into use, too. 3GPP TR 37.842 V2.1.0 defines RF requirements for active antenna system (AAS) on the base station, as of Release 13. This specification also acts as a base for the 5G era respective to the AAS.

5.10 CBRS

5.10.1 Background

The CBRS is a framework of new US FCC rules for commercial use of the 3550–3700 MHz band containing a contiguous 150 MHz block for mobile broadband services. 3GPP has extended LTE standards to support the CBRS band. It is governed by FCC to accommodate a variety of commercial uses on a shared basis with incumbent federal and nonfederal users of the band.

Access and operations will be managed by a dynamic spectrum access system like the idea used to manage television white spaces devices. The FCC stated, "Facing ever-increasing demands of wireless innovation and constrained availability of clear sources of spectrum, the Citizens Broadband Radio Service is an opportunity to add much-needed capacity through innovative sharing." [9]

Even if CBRS is not designed especially for 5G, its deployment time schedule coincides closely with the first 5G deployments and it can thus form a part of the initial 5G network infrastructure relying on the 4G technology.

CBRS provides customizable, high-capacity, private LTE networks for consumer and IoT devices. MNOs can increase capacity and coverage. Private operators get more control for special business environments such as factories, warehouses, and airports. Enterprises get seamless, reliable, and scalable LTE radio coverage. CBRS gives added value, especially in large remote sites that otherwise would lack of LTE service.

The FCC has opened 150 MHz spectrum in 3.5–3.6 GHz band. China, Japan, Korea, Europe, and other countries have allocated part or all the band for mobile broadband, and Europe has informed the band will be used in wireless 5G. Other countries are developing the concept as well [10]. ETSI is working on LSA, comparable with CBRS based on authorized shared access (ASA).

Major MNOs have been tested CBRS with special temporary licenses for capacity gain [11]. In the United States, the FCC has made the CBRS frequency available at 3.5 GHz, with 150 MHz bandwidth. The FCC Report Rulemaking 12-354, adopted by the Commission on 17 April 2015, established CBRS for shared wireless broadband use as stated in Part 96 of the Commission's rules [12, 13]. Elsewhere in the world, regulators are making similar frequencies available.

The primary user of the 3.5 GHz bands is Department of Defense, and the FCC is in the process of making the spectrum available for shared commercial use [14].

For practical deployments, the equipment must first be available. As an example, Verizon announced new vendors for CBRS in 2018. Verizon plans to offer private LTE based on CBRS, expecting handsets with CBRS band support to arrive to the commercial markets sometime around the end of 2018.

5.10.2   Use Cases for CBRS

Many of the potential CBRS applications, particularly for priority access licenses (PALs), have substantial commercial value:

- Mobile operator capacity augmentation
- Cable and MVNO (mobile virtual network operator) system augmentation
- Neutral host network for public space
- Wireless Internet service providers (WISPs), particularly in rural areas
- Enterprise LTE, etc.

CBRS allows organizations to establish their own LTE networks via spectrum sharing without purchasing spectrum license. It facilitates new use cases for consumers, enterprises, and IoT [15, 16]. It supports Automation and Industry 4.0, IoT applications in automation, remote mining and farming sites, wireless robotics in product lines, virtual-reality entertainment applications at outdoor sport events, and video surveillance. It also is adequate for mission-critical services, e.g. to monitor electricity distribution grids, and to provide ad-hoc networks for public safety agents.

There have been CBRS trials in 2018, and CBRS commercial deployment is expected latest in the beginning of 2019. The expected forerunners include tier-1 carriers, cable carriers, private LTE carriers (such as mining and transportation companies). The expected device form factors include outdoor and indoor units, smartphones, and tablets.

CBRS Alliance has quarterly meetings to review the CBRS market potential and growth. The subscriber identity module (SIM) cards are agnostic to the CBRS band and service as such, but the new and existing operators may need personalized profiles instead of the generic ones.

5.10.3   The Concept

CBRS is an add-on to the 3GPP LTE network enabling cost-effective mobile broadband in a 3.5 GHz frequency band. There is no need for the CBRS operator to purchase licensed spectrum for the shared option of the CBRS.

The shared CBRS is another variant for the unlicensed LTE, the earlier version being the LTE on unlicensed 5 GHz spectrum (LTE-U), and the LAA (Licensed Assisted Access) which is the standardized version of LTE-U as governed by 3GPP [17].

CBRS can be deployed with operators' existing packet core networks, or enterprises can deploy a private LTE network using small cells and either their own small-scale packet core or through a cloud-hosted packet core, as in the case of Nokia providing this service [18].

Universal SIM (USIM) cards are agnostic for the RF band if the wanted RAT is supported. CBRS is thus merely one variant of LTE system, and the respective USIM works as it has been designed to the LTE networks, communicating with the Home

Table 5.8 The CBRS tiers.

Tier	Users
1	Designed for incumbent users such as federal radar (occasional use e.g. in coast areas), fixed satellite services (Earth station reception), and wireless Internet service providers (ISPs).
2	Refers to the Priority Access License (PAL) holders.
3	For general authorized access (GAA) users allowing access to nonlicensed band. There is a total of 150 MHz spectrum available for the GAA-shared spectrum if the access does not interfere with PAL or incumbent users.

Subscription Server (HSS) to authenticate the user, authorize the access, and encrypt the communications. The principle of the USIM profile of the CBRS is thus the same as in any other MNO environment.

CBRS is a shared spectrum concept based on the FCC rules, providing the possibility for additional stakeholders to operate on unlicensed LTE band 48. While protecting existing national defense users' communications, FCC wants to provide an additional way to consumers for accessing wireless broadband.

Among other stakeholders, Nokia has been driving it actively, and has had a proof of concept with Qualcomm and Google. The results have showed functional high-speed video streaming in car race via CBRS infrastructure.

5.10.4 Frequency Sharing

US FCC has published updated rules for the 3550–3700 MHz spectrum allowing CBRS sharing the band by organizations. CBRS is defined for three-tiers as indicated in Table 5.8.

3GPP has specified the CBRS for LTE networks. Key technical specifications related to the radio interface and services are:

- TS 36.744, CBRS 3.5 GHz band for LTE in the United States (Release 14)
- TS 36.790, LAA (License Assisted Access)/eLAA (enhanced LAA) for the "CBRS" 3.5 GHz band in the United States

CBRS Alliance works on the development of specifications and ecosystem, facilitates the commercialization of LTE CBRS solutions, and develops a certification process for the deployments of CBRS infrastructure [11].

5.10.5 CBRS Interface

The CBRS eNodeB can be connected to the MNO's existing packet core network. It also can be deployed by enterprise as a private LTE network using small cells and either their own small-scale packet core or through a cloud-hosted packet core. One of the architectural models that can be used for offering CBRS is the multi-operator core network (MOCN) as defined by 3GPP [19].

MOCN is the 3GPP standard for radio and core; hosted clients must thus support the 3GPP core network. MOCN provides a neutral host, e.g. CBRS operator with LTE

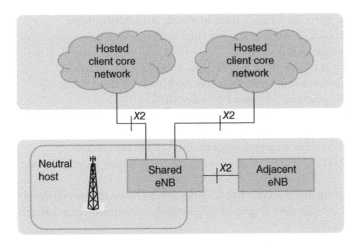

Figure 5.15 The architecture of MOCN.

eNB, connectivity to hosted client core networks (Figure 5.15). Other options include Gateway Core Network (GWCN), in which interworking is at the core network, and multiple operator RAN (MORAN), making the baseband and RF shared [20]. The hosted client must have issued users with universal integrated circuit cards (UICCs) containing 3GPP subscription info and security credentials [19].

As the CBRS is in practice an extension to the RF bands of LTE, it would not change the principle of the security the LTE network provides. LTE-based private networks thus benefit from the native SIM security, as well as emerging non-SIM options.

References

1 Ericsson (2016). *5G Radio Access*. Ericsson.
2 CBRS Alliance, "CBRS Alliance," CBRS Alliance, 2018. www.cbrsalliance.org. [Accessed 29 July 2018].
3 3GPP, "3GPP Specification series 38," 3GPP, 2018. http://www.3gpp.org/DynaReport/ 38-series.htm. [Accessed 26 July 2018].
4 Ericsson (2017). *Waveform and Numerology to Support 5G Services and Requirements*. Ericsson.
5 Lim, D.-W., Heo, S.-J., and No, J.-S. (2009). An overview of peak-to-average power ratio reduction schemes for OFDM signals. *Journal of Communications and Networks* 11 (3).
6 3GPP, "TS 38.300 V15.0.0 (2017-12)," 3GPP, 2017.
7 3GPP, "TS 38.211; Physical channels and modulation," 3GPP, 2018.
8 3GPP, "TS 38.201; NR; General description," 2018.
9 FCC, Order on Reconsideration and Second Report and Order, In the Matter of Amendment of the Commission's Rules with Regard to Commercial Operations in the 3550-3650 MHz Band, GN Docket No. 12-354, Federal Communications Commission, May 2, 2016.

10 Thinksmallcell, "Europe's plans for CBRS," Thinksmallcell, 2018. https://www
.thinksmallcell.com/Technology/europe-plans-for-cbrs-style-shared-spectrum-in-
2-3ghz-band.html. [Accessed 29 July 2018].

11 FierceWireless, "Verizon aims to deploy small cells in 3.5 GHz when practical,"
FierceWireless, 2018. https://www.fiercewireless.com/tech/verizon-aims-to-deploy-
small-cells-3-5-ghz-when-practical. [Accessed 29 July 2018].

12 Furchtgott-Roth, H. (2017). *The Potential Market Value and Consumer Surplus Value
of The Citizens Broadband Radio Service (CBRS) at 3550–3700 MHz in the United
States.* Furchtgott-Roth Economic Enterprises. IEEE http://dascongress.com/wp-
content/uploads/2018/06/The-Potential-Market-Value-and-Consumer-Surplus-Value-
of-CBRS.pdf.

13 FCC, "The FCC decisions for the 3.5 GHz band," 2018. https://www.fcc.gov/wireless/
bureau-divisions/broadband-division/35-ghz-band/35-ghz-band-citizens-broadband-
radio. [Accessed 29 July 2018].

14 RCRWireless, "Verizon's CBRS deployment plans," RCRWireless, 2018. https://
www.rcrwireless.com/20180406/carriers/verizon-names-new-cbrs-partners-tag4.
[Accessed 29 July 2018].

15 BusinessWire, "Press release of Sierra Wireless on CBRS demo at MWC 2018," BusinessWire, 2018. https://www.businesswire.com/news/home/20180225005237/en/
Sierra-Wireless-Ruckus-Networks-Showcase-Live-CBRS. [Accessed 29 July 2018.

16 Qualcomm, "Private LTE Networks. White Paper," Qualcomm, 2017.

17 IoT For All, "Unlicensed LTE Explained – LTE-U vs. LAA vs. LWA vs. Multefire:,"
IoT for All, 2018. https://www.iotforall.com/unlicensed-lte-lte-u-vs-laa-vs-lwa-vs-
multefire. [Accessed 29 July 2018].

18 Nokia, "Citizens Broadband Radio Service (CBRS): High-quality services on shared
spectrum. White Paper," Nokia, 2017.

19 Atis, "Multi Operator Core Network," Atis, 2018. https://access.atis.org/apps/group_
public/download.php/31137/ATIS-I-0000052.pdf. [Accessed 29 July 2018].

20 5G Americas, "5G Americas, shared network options.," 5G Americas, 2018. http://
www.5gamericas.org/files/4914/8193/1104/SCF191_Multi-operator_neutral_host_
small_cells.pdf. [Accessed 29 July 2018].

6

Core Network

6.1 Overview

The 5G system is designed to offer optimal resources, security, capacity, and quality for a variety of use cases. To achieve this, 3GPP has specified enhanced and new network technologies that form different service components – pillars – for servicing these use cases.

Figure 6.1 presents these main components in the 5G, referred to as *dimensions,* as can be interpreted from 3GPP SMARTER terminology presented in 3GPP TR 22.891, Release 14, which details 74 feasibility studies on new services and markets technology enablers for 5G. These dimensions support a variety of use cases while some use cases rely on multiple dimensions.

As depicted in Figure 6.1, the 5G dimensions are the enhanced mobile broadband (eMBB), critical communications (CriC), and massive Internet of Things (mIoT), supported by vehicular communications (vehicle-to-vehicle [V2V] and its variants) and network operations.

The network operations dimension provides advanced means to operate the network relying on evolved self-optimizing network (SON) principles. The automatization makes it possible to prepare the network for highly dynamic situations, as it provides the means to re-dimension the capacity and coverage on demand and perform self-healing and recover the network from faults. Especially, the network slice utilization requires highly automatized functions to cope with such a dynamic environment. In addition, it can be expected that 5G will be integrating artificial intelligence (AI) solutions and self-learning principles to enhance the performance further.

The 5G must comply with a multitude of performance criteria such as area traffic capacity, peak data rate, user's experienced data rate, spectrum efficiency, mobility, latency, connection density, and energy efficiency. The needed and offered performance depends on the type of service dimension.

As an example, the massive IoT is designed to cope with very high simultaneously communicating device density and network efficiency due to the multitude of devices functioning several years with the same battery. At the same time, e.g. the ultra-reliable low latency communications (URLLC) type of applications require the highest performance from the low latency and high mobility perspective, while it is not particularly demanding as for any other requirement parameters. The eMBB is, in turn, very demanding for all the criteria except for connection density and latency, which are somewhat intermediate in terms of importance.

5G Explained: Security and Deployment of Advanced Mobile Communications, First Edition.
Jyrki T. J. Penttinen.
© 2019 John Wiley & Sons Ltd. Published 2019 by John Wiley & Sons Ltd.

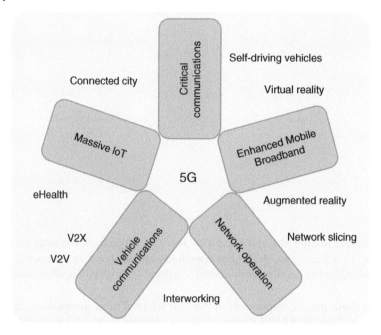

Figure 6.1 The main 5G dimensions supporting a variety of use cases as interpreted from 3GPP TR 22.891 [1].

It can be generalized that low latency is the most challenging criteria to be offered by the core network, as it needs to be as low as 1 ms for the most critical use cases. There is a variety of tiers for the 5G core deployment strategies. Some of these are:

- *Distributed data center footprint.* This refers to a model called CORD (Central Office Re-architected as Data Center). This, in turn, refers to the edge computing concept with physical assets residing closer to the core's edge, by transforming them as data centers. As the network utilization increases, the distributed data centers can support increasingly cloud-RAN and virtual-RAN hub sites, as well as virtualized core network functions.
- *Cloud-based 4G Evolved Packet Core (EPC) and 5G core.* This supports new techniques designed for the evolved 4G and 5G core networks such as separation of the user and control planes (CPs). The strict latency requirements can be achieved by applying distributed cloud infrastructure.
- *SW-defined networking.* This is needed to support the increasing edge-cloud locations for 5G, which permit high performance and secure connectivity. The centralized data centers, edge clouds, and cell sites can be thus connected via wide-area software defined networking (SDN) in an efficient way as it provides adequate performance and level of routing, security, redundancy, and orchestration.
- *Network slicing.* This concept optimizes 5G core services per use case. Network slicing refers to the concept where the virtual network functions (VNF) are formed by service, which results in optimized processing paths for the data delivery over the whole network. Furthermore, the slices are isolated from each other, including the security aspects.

- *Network operation automation.* This concept makes it possible to take new services in use efficiently and fast. The impact can be monitored and analyzed, and it is possible to modify and scale the services in a highly dynamic way.

The 5G core network is standardized to define a functional architecture in such a way that the implementation and replacement of further technologies is possible, and the 5G interfaces support much more fluently through multivendor scenarios than in previous generations. Furthermore, the 5G architecture provides separate scaling of the user plane (UP) and control plane (CP) functionality, which facilitates the deployment of the UP and CP independently. Furthermore, 5G supports authentication for both 3GPP and non-3GPP identities, which makes it possible to fluently use e.g. Wi-Fi networks as a part of the communications. Also, 5G allows the utilization of different network configurations via respective network slices.

The most essential new core network concepts in 5G are thus the separation of UP and CP, network slicing, and service-based architecture (SBA).

6.2 Preparing the Core for 5G

6.2.1 Cloud Concept

The 5G system will change the philosophy of earlier mobile communication networks. One of the concepts indicating this evolution is the introduction of the cloud concept. This is a result of increasing the intelligence of the 5G network, which, in turn, enables the user devices and network equipment to communicate efficiently via optimized transport and core functionalities.

In practice, this means that the cloud concept is utilized for enabling intelligent service awareness, which, in turn, optimizes the connectivity, latency, and other essential performance characteristics. Thus, 5G networks will be flexible and scalable to offer advanced services to the end users, which can be done by virtualizing the network functions.

As a result, the role of data centers will increase considerably, along with 5G network deployment. They enable highly dynamic, scalable, and flexible offering of capacity, whether the respective connected device is a tiny IoT sensor or advanced equipment capable of supporting high-speed data for virtual reality purposes.

In addition to centralized clouds hosted in data centers, 5G application can also reside in network's edge, closer to the end user. These applications can be hosted in mobile edge computing nodes, referred to as cloudlets.

6.2.2 Data Center as a Base for 5G Architecture

5G will require increased cloud computing to fulfill the strict performance criteria for the advanced communications. Data centers must be upgraded thus to satisfy this need for 5G traffic. The elemental requirements are related to increased central processing unit (CPU) performance and storage capacity, which requires increased number of servers, enhanced cooling of the physical equipment, and more space for housing the respective racks.

Most remarkably, data centers require fundamental architectural renewal from the previous decentralized mobile network model to better serve the centralized processing of the network functionalities. As a result, instead of specific equipment for the mobile system purposes, there will be cloud solutions both centralized as well as in edge regions.

The network functions (NFs) are increasingly executed in these clouds in virtualized software environment instead of the local HW-processing. This principle could potentially evolve in such a depth that major part of the base station processing takes place in clouds and is handed over back to the local processing of the original base stations merely in events of heavy cloud congestion.

There is a trend for relying on Open Compute Project (OCP) in so called hyperscale data center operators. This concept can be assumed to be adopted increasingly also to telecommunication operations [2]. According to this source, various telecommunications companies investigated proof of concepts (PoCs) by applying Open Compute in their data centers. The significance of the Open Compute is expected to increase considerably in 2018 and beyond, to serve as telecommunications cloud for OCP. In fact, telecommunications may turn out to be the driving force for such development of OCP, accompanied by the Telecom Infra Project (TIP).

According to [3], the trend of cloud business applications, including 5G and IoT in general, can lead to the Internet becoming increasingly distributed. This means that more data centers need to be located closer to the users. The mayor benefits of these compact edge data centers include the ability to reduce costs and latency. This, in turn, leads into a hierarchical architecture consisting of centralized data center and several connected edge clouds as can be interpreted from [4]. The architecture is summarized in Figure 6.2 as interpreted from [3].

Edge computing is thus an elemental part of the distributed data center infrastructure, ensuring that the computing and storage resources are optimally close to the users [4].

The benefits of edge computing include ultra-low latency, reduction/offloading of core network (backhaul) traffic, and in-network data processing. Edge computing facilitates the optimal performance so that latency-sensitive applications benefit from the concept. Some examples of these special environments are autonomous devices, including self-driving vehicles, drones, and robots, as well as applications offering immersive user

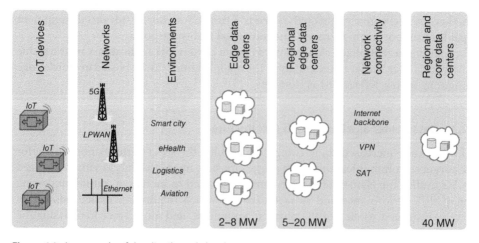

Figure 6.2 An example of the distributed cloud concept.

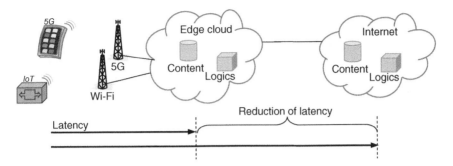

Figure 6.3 Edge computing reduces latency.

experience such as virtual reality and augmented reality solutions with highly interactive interfaces requiring practically real-time response times (Figure 6.3).

The distribution of network functions can be divided, e.g. into common functions such as slice control, distributed data functions such as data forwarding and quality of service (QoS) control, and slice-specific control functions such as session/mobility control and policy management while the transport network delivers the data and control information between the network functions.

The main reason for the utilization of the edge data centers is the significantly lower latency and faster delivery of the content. Also, the cost of the transport can be reduced significantly.

It can be assumed that much of the IoT traffic will be generated locally due to the characteristics of the devices; they typically are not too mobile but stay in their special environments such as factories, whereas at the other extreme, the massive IoT adapted to the V2X environment can be very mobile.

6.2.3 Service-Level Assurance

The service-level agreement (SLA) is typically used between a service operator and respective customer relying on the jointly agreed QoS level. The SLA can be applied in many environments as a standard agreement type, including mobile network operators (MNOs) promising certain service availability for the customers, making the agreement with national regulator, and outsourced service provider of mobile network.

Compliance with the SLA can be monitored and recorded based on events that cause outage to the service. These cases may occur if e.g. a server's hardware fails or experiences software issues. One example of these cases is the corruption of the database, which may be caused by faulty SW (pointers failing) or HW (hard disk or memory chip failing).

If such an event triggers degradation of the expected quality of service, the SLA monitoring registers the start time of the problem, which can be e.g. raising of service request by the customer. It can be equally automatized based on performance monitoring (PM) or fault management (FM) reporting, or some other appropriate service-level reporting (SLR) method. The SLA monitoring also registers the resolution time stamp that can be manual confirmation of the solved service ticket, or again, automatic PM or FM reporting.

The traditional, standalone network architecture with dedicated HW and their respective SW for network functions and services requires some of the well-known strategies for the sufficiently fast recovery of the faults, and in the worst case, performing a complete disaster recovery.

Some of the means to do this in practice are, e.g. active-passive redundancy, which means that there is one operational server setup while another is in nonoperational mode ready to take over the failing system after certain ramp-up time. Active-active setup runs two parallel servers reducing downtime in case the master server system fails. Geo-redundancy combined with active-active strategy provides very high availability. It means that the service is guaranteed by running the operations in two physically separated servers in a parallel fashion; if one machine fails, the operations are synchronized so that the other equipment can take over the master role and continue the operations with only insignificant delay, if any, in service continuum. In this way, if one of the physical sites is damaged, e.g. due to natural disaster, there is no service downtime as the counterparty is safely located far enough.

As a practical example of the SLA, referencing [5], Google Cloud Platform has license agreement ensuring that the covered service will provide a monthly uptime percentage to customer of at least 99.99%. This value is referred to as the service level objective (SLO), or operating level agreement (OLA). Furthermore, if Google does not meet this agreed SLO, assuming the customer meets its own obligations, there will be a financial credit triggered in favor of the customer. This type of financial compensation is typical in SLA, although it depends solely on the negotiating parties, as well as the actual uptime percentage and the monitoring period of it.

The monthly uptime percentage t_u refers to a value that can be calculated in as follows:

$$t_u = \frac{t_{tot} - t_d}{t_{tot}} \tag{6.1}$$

In this equation, t_u is the time of total minutes in a period of a single month, and t_d is the number of minutes of downtime counting all downtime periods in the month. Even the equation does not rely on advanced mathematics, it is highly useful and adapted to operations. Figure 6.4 presents example of the service uptime values (%) and respective minutes the service is impacted. If the SLA assumes maximum service outage of 5% per month, i.e. minimum 95% uptime during the observed SLA period, it converts to a maximum of 36.5 minutes.

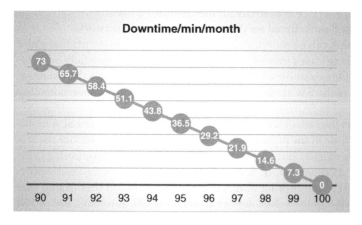

Figure 6.4 Examples of the service outage time. The *x*-axis represents the uptime in %.

In 5G, the SLA can be assumed to be much more demanding, especially for the CriC-type of use cases. Some 5G environments require extremely high reliability, such as the applications relying on the URLLC. In some cases, the service uptime could be required to be as high as 99.999% (sometimes referred to as "nine fives"), or more. This, in turn, requires high availability of the network elements, which are largely housed at data centers, with protection mechanisms based on geo-redundancy. Logically, the more demanding the availability is, the more expensive the underlying solution is.

In 5G, one way to meet the demanding performance requirements is to rely on network slices that run in end-to-end chain, including radio and core networks as well as application logic [6]. In order to make this happen, a service and resource orchestration is needed across the main functional domains. Such orchestration creates other issues, though, because each of these domains have their own control system:

- Transport network can rely on SDN for traffic segmentation and policy selection.
- Radio access network (RAN) can rely on radio network controller or base station for management of packet scheduling.
- Cloud core network can rely on network functions virtualization (NFV) orchestrator.

In normal operations, the service assurance of the 5G network slices needs to happen automatically without human intervention except for special cases. The above-mentioned domains need thus service layer's assistance in the inter-domain coordination as can be seen in Figure 6.5.

Each 5G network slice equals in practice a separate, self-contained network for specific use cases. In 5G, each slice can thus separate SLA, which is determined per each customer with the service provider. This is the reason why the service management and service assurance need to be divided into each network slice separately.

It is also worth mentioning that the lifetime of some of the slices may be quite short, which can happen e.g. when it is used for some special event only for a limited time. Thus, the SLA is accordingly of short-term, and it considers the initial and termination phase of the slice setup. The challenge of this case is the fact that the same resource layer is utilized as a basis for multiple slices, and the assurance of the adequate resources per

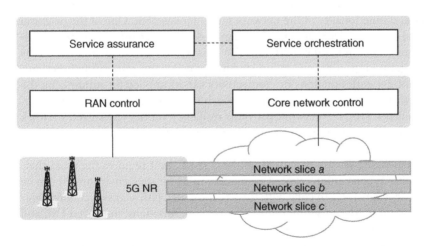

Figure 6.5 The principle of the service assurance for 5G network slicing.

slice may not be possible to achieve in some of these special cases due to unforeseen peak capacity demand.

The solution for this issue is the deployment of adequate service assurance tool that can correlate service experience with the respective resource utilization. In optimal setup, this ensures that a single slice does not consume more resources than absolutely needed to fulfill the respective SLA, considering the performance and capacity. The strategy for such balancing depends on the use case, so the service priority for, e.g. a set of noncritical mIoT sensors is lower than for applications requiring ultra-low latency. This type of balancing is known as contextual service assurance between the network management and orchestration plane (NMO), and the slice layer [6].

Ref. [7] has identified examples of the network slice utilization in practice. One of them is automotive slice for connected vehicles. This use case requires a network that can deliver simultaneously high bit-rate data for the use of in-car entertainment, as well as data of URLLC category for autonomous driving of the vehicle, accompanied by telemetry data from the vehicle's sensors and device to device communication. In this type of environments, it is of utmost importance to ensure the service continuity in all the situations, including the internal roaming. For more information on commercial solutions, please refer, e.g. to [8, 9].

6.2.4 Resiliency

As it relates to the telecommunications cloud environment, resiliency refers to the means to provide and maintain acceptable service when faults occur. There are several technical means to manage the resiliency from which examples are FM, resource redundancy, and configuring edge cloud resources, so that they can take over a significant part of the tasks, minimizing the connectivity toward the central cloud. Another way is to enhance resiliency of key network functions. Ref. [10] summarizes feasibility of resilience in 5G architectures, combining the closely related topic of security into the analysis.

Security and resilience are mutually dependent. Many security threats may thus affect the resilience and functionality of the network fault management. One of the practical cases is a denial of service (DoS) attack, which can lead to reduced utilization or unavailability of network functions and underlying HW. This type of threat can be detected by FM and the negative impact may be minimized by the respective restoration functions. Also, redundant network functions or connectivity can be applied for minimizing the impact. Nevertheless, the downside of this approach is that it may lead to lower redundancy and resilience [10]. 5G network provides resiliency via the virtualized network architecture (based on telecommunications cloud), and physical HW-instances such as selected functions of RAN.

The most challenging service type of 5G is the URLLC, as it requires the most challenging reliability. Typical uninterrupted availability value for such environment can be expected to be 99.999% [11]. To cope with the situation, Ref. [10] suggests countermeasures via redundant hardware setup, redundant network functions, improved FM, and in-built resilience of network functions.

The usability of FM for resilience is based on its ability to monitor changes in the network performance, and to provide recovery suggestions to the determined causes of issues. The FM can include root-cause analysis for identifying and detailing the issue, and for isolating it to minimize the negative impact on the network performance [12].

The benefit of virtualized network is the removal of the strict dependency between HW and SW per dedicated element performing its designed functions. The new service architecture thus changes the philosophy of the FM. As identified in [10], virtualized network has three deployment layers, which are network function and service logic, virtual infrastructure, and physical infrastructure. The network function logics contains network functions (NF), which in legacy infrastructure belonged to each dedicated machine for respective task. The service architecture has virtualized infrastructure, which can include, e.g. virtual machines (VMs) and containers. The underlying physical infrastructure contains commercial off-the-shelf (COTS) servers and data storage components in a virtual environment, which serve as a common base for the network functions.

6.2.5 Redundancy

To ensure the lowest outage time, the network needs to be more redundant. There are different configuration models for optimal redundancy solutions; the highest availability is logically much more expensive than basic protection, so it's matter of balancing the need and cost.

Specifically, for data centers, the redundancy needs to be considered basing on the number of physically separated power lines, capacity of unbreakable power system (UPS), wake-up time for possible diesel generator, and redundancy for the actual telecommunications lines. The potential reasons for the failures in power loss and connectivity include equipment failure, SW failure, natural, or other type of disaster. The latter may be caused accidentally or even deliberately, making the malicious human aspect relevant as one of the evaluation criteria.

The efforts for minimizing such outages are covered by certification of data centers. There is a variety of certification schemes in commercial market, and the compliance of requirements of any of entities counting for good reputation gives added business value for the data center operators. The certification scheme may include the physical assurance of the functioning as well as the security aspects for confidentiality of data. Some practical examples of the requirements are the resistance against earthquakes and storms (which can be coped with underground premises and flexible floors), and physical shielding of the equipment premises (e.g. fingerprint reader, multiple locked doors, and means to ensure that minimum two representatives must be present at the same time in the premises).

The means for minimizing the outage time (i.e. maximizing the uptime of data center), must be planned according to the need. Some of the typically applied configurations are active-passive and active-active.

In addition to the core, 5G also renews RAN by introducing the new radio (NR). The most challenging reliability requirements for the access point of view are for the URLLC, while the eMBB provides high data speed, and mIoT is the enabler for mIoT space [10]. The URLLC may require in some cases uptime values close to 99.999%. This requires new technical solutions for maintaining such a challenging value. Some examples are data duplication and network coding.

As the parallel systems are ensuring the fluent continuum, regardless of the time to recovery from disaster, the critical topic remains that the system can still fail if a single point fails. With redundant telecommunication infrastructure, if there is a unique, sole point for e.g. feeding the power for both master and parallel systems, and if that point is compromised, the active secondary system would not help until the single point of failure is repaired.

In addition to the system and its network functions, another important aspect is data storage. In high-availability scenarios, the data repository must be duplicated (or multiplied). Again, the most reliable way of doing this is to ensure that the physical locations of such backups are sufficiently separated in case of, e.g. fire of a complete building.

6.2.6 Recovery Classes

There are different classes and tiers of redundancy based on disaster recovery schemes for this. As an example, in the 1980s, there was a SHARE Technical Steering Committee (classes 0–7) tier criteria developed for data backup levels. Nevertheless, up-to-date example of the solutions suitable for current operations is the Computer Network Technologies (CNT) scheme, which categorizes recovery time objectives (RTOs) and recovery point objectives (RPOs) into different classes (classes 1–4). That is currently adequate to categorize the data center recovery, too.

Some datacenters specialize in disaster recovery. These may consist of networks of data centers and disaster recovery sites that are designed for delivering recovery as a service ensuring adequate availability, performance, and resiliency [13].

One of the typical representatives is based on CNT, which has developed category schemes for RPO and RTO. The RPO is the time in the past it has taken to recover. RTO, in turn, refers to the time in future at which the system is again functional. The classes of CNT scheme are summarized in Table 6.1.

In other words, RTO refers to the time that it takes from outage occurrence to execute a failover recovery and resumption of services. RPO, in turn, refers to the amount of data that was lost due to the outage.

6.2.6.1 Active-Passive Configuration

Active-passive configuration refers to a system that has a backup system running in a parallel fashion, but it is not available for an immediate takeover if the primary system fails, as it is in a standby mode. The time for such handover procedure depends on the level of the redundancy, which, in turn, contributes mostly to the total outage time. Practical example of system with active-passive configuration is Structured Query Language (SQL) cluster, where at least one of the servers is reserved as a standby to accept failover of an SQL instance from the active server in case of such occurrence.

Table 6.1 The criteria for CNT classes 1–4.

Class	RTO (recovery)	RPO (last full backup)
1	72 h–1 wk	<1 wk
2	8–72 h	<24 h
3	<8 h	<15 min
4	0	<0.1 s

Active-passive configuration is adequate for less-critical applications. This type of data can thus tolerate longer recovery times and lost data. To measure the continuity, the RTO and RPO as described earlier in this chapter can be used.

Typical aspects to be considered in active-passive configuration are:

- Availability of the system and amount of lost data improve along with moving from active-passive to an active-active architecture; equally, the improvement happens when moving from asynchronous to synchronous data replication solutions.
- The asynchronous active-passive scenario offers typically recovery time of seconds to hours and recovery of data in seconds to minutes. This scenario is popular to balance the expense and value of the lost data and time while it is not adequate in a mission-critical environment or even less-critical use cases, according to current quality expectations of the end users.
- Asynchronous active-active solution outperforms the above-mentioned asynchronous active-passive, offering typical recovery values of milliseconds to seconds, and its data loss performance is better, too.
- The synchronous replication of the above-mentioned architectures does not change the availability values compared to asynchronous mode, yet the advantages may be the nonexistent data loss and data collisions in an active-active scenario.

A backup node is required when aiming to achieve higher service availability levels. In an active-passive configuration, the passive system is normally in idle mode for update procedures although applications may also be active in read-only mode in the standby server. Furthermore, the standby database may be actively used for query and reporting [14].

6.2.6.2 Active-Active Configuration

The principle of active-active configuration is that the critical functions of the system's HW and SW are executing the same data flows in a parallel fashion. If one of the system entities fail, the duplicated entity takes over the respective communications. Practical example of system with active-active configuration is SQL cluster where all servers participating in an SQL server cluster are running at least one SQL instance. In case of active-active data centers, there are two (or more) datacenters from which all sites can serve an application at any time. This configuration is adequate for business-critical applications by synchronizing the respective databases.

The number of total entities in active-active system is 2. Technically, the number of entities in active-active configuration can be even higher, and the parallel entities can be anything from one or more. The more redundant entities there are, the lower the probability of a total service outage, respectively. The downsize of multiple entities is that the cost of such solution may turn out to be very expensive compared to the benefits of minimizing the outages, and the synchronization may get unpractically complicated.

Ref. [15] presents a practical example on the configuring of an active-active layer 3 cluster. It should be noted that in this example, the active-active deployment refers only to the active-active mode of the user plane while the control plane is in active-passive mode. This means that a single control plane controls both servers as a single logical device. In case of failure, the control plain takes over the remaining server while the data plane can make failover independently of the control plane.

The active-active configuration, as well as active-passive scenario, can be enhanced further by applying geo-redundancy. It means that the servers are separated physically to locations sufficiently far away from each other. The benefit of such solution is that if there is a vast physical damage in one site, the system taking over the failing one is not impacted. The georedundancy can be applied in fact in a straightforward way to the edge computing scenarios where locally consumed contents is moved closer to the customers, and thus the respective cloud equipment is separated physically from the central cloud.

6.2.7 Connectivity Service Network

There will be a variety of services offered based on 5G infrastructure. As for the customer's credentials, i.e. subscriber identity module (SIM) profile management, one of the commercial examples is connectivity service network (CSN) by Giesecke+Devirient. It has been designed to support in a central role as a service enabler, particularly with the explosion of IoT and the launch of 5G [16].

The CSN provides means for MNOs to manage SIMs, devices, and connections with high availability and secure connections. The solution is horizontally scalable, which means that other services can be added without impacts. The high availability fulfills the requirements for mission-critical applications and eSIM management. It integrates an intelligent monitoring system for performance and load success rate on universal integrated circuit cards (UICCs) and devices with polling mechanisms automatic update [17].

6.2.8 Network as a Service

One of the key documents for cloud environment is the International Telecommunication Union (ITU) Recommendation ITU-T Y.3515, which provides Network as a Service (NaaS) functional architecture [18].

The Recommendation ITU-T Y.3500 defines NaaS as a cloud service category in which the capability provided to the cloud service customer is transport connectivity and related network capabilities. Concretely, cloud service provider is the party that makes cloud services available for cloud service customers. The respective arrangements involve typically SLA, which is documented agreement between the service provider and customer that identifies services and service targets. It is also possible to establish a SLA between the service provider and a supplier, an internal group or a customer acting as a supplier.

It specifies functionalities, functional components, and reference points for the operation support system (OSS). It also details the mapping between functionalities and functional requirements of NaaS, relationship between the NaaS functional architecture and SDN and illustrated usage of SDN and NFV in support of the NaaS functional architecture.

There is also a set of ITU recommendations highly relevant for the cloud environment, and the housing of 5G functions are dependent on many of these directly or indirectly:

- ITU-T X.1601 (2015), Security framework for cloud computing
- ITU-T Y.2320 (2015), Requirements for virtualization of control network entities in next generation network evolution

- ITU-T Y.3300 (2014), Framework of SDN
- ITU-T Y.3302 (2017), Functional architecture of SDN
- ITU-T Y.3500 (2014), Information technology – Cloud computing – Overview and vocabulary (aligned with ISO/IEC 17788:2014)
- ITU-T Y.3501 (2016), Cloud computing – Framework and high-level requirements (aligned with ISO/IEC 17789:2014)
- ITU-T Y.3502 (2014), Information technology – Cloud computing – Reference architecture
- ITU-T Y.3512 (2014), Cloud computing – Functional requirements of NaaS
- ITU-T Y.3521/M.3070 (2016), Overview of end-to-end cloud computing management
- ITU-T Y.3522 (2016), End-to-end cloud service lifecycle management requirements

Based on these references, focus on the ITU-T Y.3512, the NaaS services consist of NaaS application, platform, and connectivity services. Some examples of the NaaS application and NaaS platform services include virtual Internet protocol multimedia subsystem (vIMS), virtual evolved packet core (vEPC) and virtual content delivery network (vCDN) while NaaS connectivity services can be, e.g. virtual private network (VPN) and bandwidth on demand (Figures 6.6 and 6.7).

These NaaS services rely on network services and network functions. NaaS Cloud Service Provider (CSP) provides them on-demand-basis to the NaaS Cloud Service Category (CSC). If NaaS CSP deploys the network functions as software, they are representatives of VNF. Another option is to deploy them coupled in unique, separate software and hardware systems. In that scenario, the network functions are representatives of *physical network functions* (PNF). Both scenarios are applicable to the 5G architecture models, although the full benefits and optimal performance can be obtained by the virtualized environment.

Figure 6.8 depicts the principle of NaaS CSP as interpreted from ITU-T Y.3515 (July 2017).

Figure 6.6 The balancing of resiliency and security as depicted in [10].

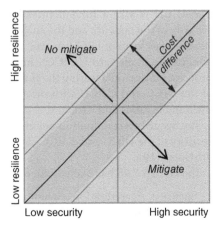

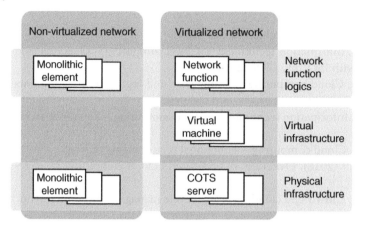

Figure 6.7 The difference between legacy and virtualized networks.

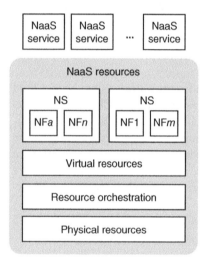

Figure 6.8 Principle of NaaS CSP as defined in [18]. The virtual resources refer to, e.g. computing resources, storage, and networking.

6.2.9 Security Certification of 5G SIM and Subscription Management

The UICC is still a fundamental key component in 5G systems for ensuring the security of the network and the subscriber's credentials, services, and transactions. The "traditional" form of UICC, as well as the embedded variant of it, embedded universal integrated circuit card (eUICC), must be protected as the supplier environment and processes for manufacturing them are considered. This applies to ensuring the integrity of the UICC and eUICC with or without remote provisioning capabilities, as well as their applications and data.

The Security Accreditation Scheme (SAS) of GSM Association (GSMA) assesses the security of the UICC and eUICC suppliers and eSIM subscription management service providers [19].

The GSMA SAS has been divided into two schemes:

1. Security Accreditation Scheme-Universal Integrated Circuit Card Production (SAS-UP) refers to UICC production. This scheme has been deployed since 2000.

It means that the production sites and processes of UICC manufacturers undergo a thorough security audit. The compliant sires are granted a security accreditation for one-year period, which is extended to two additional years upon each successful renewal. SAS for UICC production is described in GSMA documents FS.04 SAS-UP Standard v8 and FS.05 SAS-UP Methodology v7.

2. SAS-SM refers to subscription management. This scheme has been adapted to create confidence of the remote provisioning for embedded SIM products, extending the SAS-UP model. SAS for subscription management (SM) is described in documents SAS for Subscription Management FS.08 SAS-SM Standard v3 and FS.09 SAS-SM Methodology v4.

In addition to the above-mentioned documentation, GSMA also has produced a common document applicable to both schemes under the title FS.17 SAS Consolidated Security Requirements v2.

6.2.10 Security Certification of Data Centers

The current telecom and overall data service business trend is to rely increasingly on virtualized data center networking and SDNs. In the professional environment, certificates are applied both for the personnel operating the data centers as well as for the premises; the latter is getting increasingly critical as the data will include also more sensitive information. 5G architecture includes critical elements such as data repository and authentication which are housed by default within the MNO cloud infrastructure; nevertheless, there are no technical restrictions to rely on functional elements managed by third-service providers for such tasks.

The personnel of data centers need to master management of cloud data, considering security aspects. The certification of data center personnel requires thus typical knowledge of networking technologies, including virtualization and cloud technologies ensuring the adequate safety aspects. There are many certificates designed for the experts in this domain.

There exists a variety of certificates for the data centers, too. For the security of the data centers, some of the commonly recognized entities managing the certificates include Health Insurance Portability and Accountability Act (HIPAA), Payment Card Industry (PCI), Uptime Institute, and Colocation America [20].

The certified data center brings additional confidence for the parties relying on the center. For the critical functions, including the ones 5G requires, there must be a guarantee for the security. It is not solely a matter of business partners trusting each other but the statutory certifications are required by law. An example of this is the HIPAA and requires auditing prior to the operations. The standard certifications, on the other hand, are a set of requirements of certain authority entities dictating criteria for performance operations. Unlike the stationary certification, these tier standards are not required by law; nevertheless, they do have an important role in data center business. The standard certified refers to other type of principles as the certification is based on data center owner compliance plans considering expenses and specific needs of the potential customers. This type of certification process is more a matter of reputation, which may be challenging to prove if the data center is about to start operations.

6.2.10.1 Overall Certificates

The standards designed by US HIPAA are based on audit system to ensure that the evaluated data center facility is compatible with rigorous requirements of the federal regulation. The audits are carried out in practice by independent inspectors. Originally, the US HIPAA certification process was derived from the safe transfer and storage of Protected Health Information (PHI); there is a total of 19 HIPAA standards for the healthcare industry.

Another certificate is of the PCI, which manages Data Security Standard (DSS). It is designed to protect consumers' transactions based on credit cards.

A third example is the Uptime Institute, which is based on a four-tier ranking system to benchmark the reliability of data centers.

6.2.10.2 Data Security Certificates

The CPA (Certified Public Accountants) has set up a SSAE 16 certificate (Statement on Standards for Attestation Engagements 16). It is in practice a set of guidelines for the level of controls at a service organization. The guidelines are applied to safe storage of data in the servers, and for the secure transfer of the data within and between data centers and external entities.

Another example of the security certificates is Service Organization Control (SOC) reporting framework. It contains three reporting standards, which are SOC 1, SOC 2, and SOC 3. The SOC refers to a reporting standard for a business' financial reports.

6.3 5G Core Network Elements

The 5G core network is based on cloud concept, and the most efficient forms of the core will take advance of network slicing and service-based architecture. The deployment of these advanced features enhances the end-user's perception of the QoS as the network develops.

For the incumbent MNO, the first step for practical core deployment is to take advantage of the already existing 4G infrastructure. This happens by using virtual 4G core, which is prepared for the transition, i.e. the already existing infrastructure can be prepared to provide a base for the "5G-ready" networks. Then, relying on this base, it is possible to deploy new software functionalities such as network slicing, cloud-based VNF, SDN, and distributed user-plane and automatization. The existing network can be thus optimized and scaled up to support increasing 5G-type of use cases.

The 5G network elements are depicted in Figure 6.9. While some of the elements are mandatory, others are optional. Furthermore, some elements can be collocated physically into other elements. In 5G, the previous Long-Term Evolution (LTE) elements and their functions have been reorganized, and new network functions are introduced. The service-based architecture reorganizes also the control plane so that instead of Mobility Management Entity (MME), S-GW and P-GW as they have defined in EPC, their functionality has been divided into AMF (Access and Mobility Management Function) and SMF (Session Management Function). Along with the network slicing concept, which is new in 5G, a single user equipment (UE) can have one or more SMFs associated at a time, each representing a network slice.

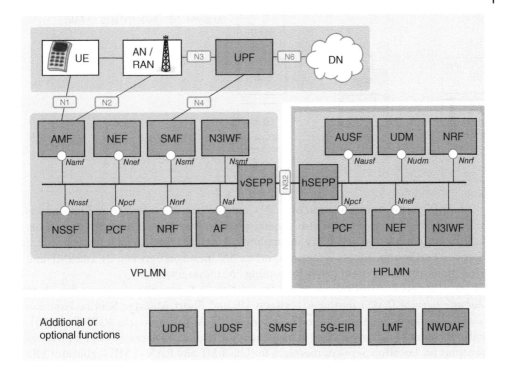

Figure 6.9 The 5G network function elements.

The following sections present the core architecture of some of the key elements of 5G and how they are positioned in the core network as interpreted from 3GPP TS 23.501 [21].

6.3.1 5G-EIR

The 5G Equipment Identity Register (5G-EIR) is an evolution of LTE EIR. As has been the principle in previous generations, it is also optional functionality in 5G network. The task of 5G-EIR is to check whether PEI has been blacklisted.

6.3.2 AF

The application function works in application layer of 5G, and it is comparable with the LTE access stratum (AS) and gsmSCF. It interacts with the 3GPP core network to provide services such as application influence on traffic routing, access to Network Exposure Function (NEF), and interaction with the policy framework. Depending on each deployment scenario, AF that the operator trusts may interact with relevant NFs while the nontrusted AF rely on external exposure framework via NEF for the interaction with NF.

6.3.3 AMF

The 5G AMF replaces the LTE MME. It is defined in 3GPP TS 23.501. AMF has a multitude of tasks and it interfaces with various functional elements as depicted in Figure 6.10.

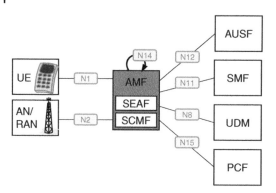

Figure 6.10 The interfaces of AMF.

The AMF performs termination of RAN Control Plane *N2*-interface and non-access stratum (NAS) *N1*-interface. It also manages NAS ciphering and integrity protection, registration, connection, reliability and mobility. It includes the Lawful Interception (*LI*) interface. Furthermore, it provides transport for SM messages between UE and SMF, and it functions as a transparent proxy for routing SM messages.

It does access authentication and authorization and provides transport for Short Message Service (SMS) messages between UE and Short Message Service Function (SMSF). It embeds Security Anchor Functionality (SEAF) according to 3GPP TS 33.501. It also includes Location Services management for regulatory services. And it provides transport for Location Services messages for UE–LMF and RAN–LMF. It allocates EPS Bearer ID for EPS-interworking, and it manages UE mobility event notification.

6.3.4 AUSF

The 5G Authentication Server Function (AUSF) replaces the LTE MME/AAA. It supports authentication for 3GPP access and untrusted non-3GPP access a TS 33.501. It has interfaces toward AMF and Unified Data Management (UDM) as depicted in Figure 6.11.

6.3.5 LMF

The LMF can determinate the location for a UE. It can obtain downlink location measurements or estimate of the location from the UE, and uplink location measurements from the NG RAN. It can also obtain non-UE associated assistance data from the NG RAN. Figure 6.12 depicts the interfaces of LMF.

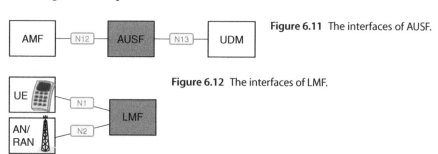

Figure 6.11 The interfaces of AUSF.

Figure 6.12 The interfaces of LMF.

6.3.6 N3IWF

The Non-3GPP Interworking Function (N3IWF) refers to the N3IWFs, which take place when untrusted access is used. In this case, the N3IWF establishes IPsec tunnel with the UE. The N3IWF uses the IKEv2/IPsec protocols with the UE over *NWu* and relays over *N2* information to authenticate and authorize the access of the UE with the 5G Core Network. The N3IWF has termination points of *N2* and *N3* with the 5G core. It also relays uplink and downlink control-plane NAS signaling of *N1* between the UE and AMF, and it handles *N2* signaling from SMF and relayed by AMF related to PDU Sessions and QoS. It establishes IPsec Security Association (IPsec SA) for PDU Session traffic.

The N3IWF relays uplink and downlink user-plane packets between the UE and User Plane Function (UPF). This task includes the following: Packet de-capsulation and encapsulation for IPSec and *N3* tunneling; QoS enforcement for *N3* packet marking based on the information received from *N2*; *N3* user plane packet marking in the uplink, and support in AMF selection. In addition, N3IWF can act as a local mobility anchor within untrusted non-3GPP access networks using MOBIKE as described in IETF RFC 4555. Among other benefits, MOBIKE enables a remote access VPN user to move from one address to another without reestablishing all security associations with the VPN gateway [22] (Figure 6.13).

6.3.7 NEF

The 5G NEF is an evolution of Service Capability Exposure Function (SCEF) and application programming interface (API) layer of LTE. In 5G, it assists in storing and retrieving exposed capabilities and events of NFs via *Nudr* interface with unified data repository (UDR). This information is thus shared between NFs in 5G network. The NEF can also expose securely this information to entity residing outside of the 3GPP network, such as third-party and edge-computing entities. As depicted in Figure 6.14, the NEF can access only the UDR residing within the same public land mobile network (PLMN). For the external exposure, NEF can communicate with the respective entities via Common API Framework (CAPIF) and respective API. This procedure is detailed in 3GPP TS 23.222.

Via the NEF, 3GPP network functions can retrieve information needed for efficiently managing the connections. An example of such information is the capabilities and expected behavior of the UE. The NEF is also capable of translating information such as AF service identifier and 5G core-related information, e.g. Data Network Name (DNN) and Network Slice Selection Assistance Information (SNSSAI). One of the important tasks of NEF is to mask sensitive information to external AFs as per the MNO policy rules.

Figure 6.13 The interfaces of N3IWF.

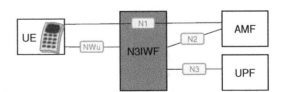

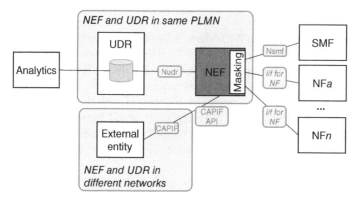

Figure 6.14 The interfaces of NEF.

The NEF stores the available information retrieved from NFs into UDR in structured data format. This information can be exposed to other NFs via NEF or be utilized for other tasks such as data analytics.

The NEF may also support a PFDF (Packet Flow Descriptions Function). In 5G, this could be the case, e.g. in *Sx* PFD management procedure to provision PFDs for one or more Application Identifiers in the UP function (work item of Sponsored Data Connectivity Improvements, SDCI) [23]. The PFDF is a repository that stores packet flow descriptions (PFDs). They can be added, updated and removed by the SCS/AS via the SCEF.

As an example, an IoT platform serving business customer may want to add a new PFD allowing sensor traffic to pass to a new server. In this example, the PFD could consist of PFD-ID, Server-IP address, protocol, and port number. This, in turn, needs to be enforced and handled by PCEF (which can be P-GW in the case of 4G). PFDF thus enables third-party AS to provide, modify, and remove PFDs via the SCEF into the MNO network. PFDF, in return, may store the PFDs or provision it to PCEF via JSON interface (*Gw*) [24].

In 5G, the PFDF in the NEF may store and retrieve PFDs in the UDR. The NEF provides these PFDs to the SMF on pull request. The PFDs can also be delivered to SMF triggered by NEF push request. This procedure is detailed in 3GPP TS 23.503.

6.3.8 NRF

The 5G NF Repository Function is a part of the evolution of DNS utilized in LTE. In 5G, it supports service discovery function for providing the information of the discovered NF instances to the requesting NF instance. One of the tasks of the NRF is thus to maintain the NF profile and respective services of available NF instances. The UDR profiles are standardized, and in such scenario the NF profile contains the following: NF instance ID; NF type; PLMN ID; network slice identifiers such as SNSSAI and NSI ID; FQDN or IP address of each NF; information on NF capacity; authorization information related to the NF; supported service names; endpoint address of each instance of supported services; and identification of stored data.

The applicability of this information as for other NF profiles is left up to implementation: notification endpoint for NF service notifications; routing ID of Subscriber

Concealed Identifier (SUCI); A set of GUAMIs and TAIs if AMF is involved; UDM Group ID and if UDM is involved; UDR Group ID if UDR is involved; and AUSF Group ID if AUSF is involved. The NRF is capable of mapping between abovementioned UDM, UDR, and AUSF Group ID and Subscriber Permanent Identifier (SUPIs) to facilitate the discovery of UDM, UDR, and AUSF by SUPI.

Furthermore, multiple NRFs may be deployed in multitudes of networks: NRF in the Visited PLMN is referred to as vNRF, and NRF in the Home PLMN is known as the hNRF, which is referenced by the vNRF via the *N27* interface.

6.3.9 NSSF

The 5G Network Slice Selection Function is a new function to support the 5G-specific network slicing concept. The NSSF selects needed set of network slice instances serving the UE. It also determinates the allowed and configured NSSAIs and if necessary, it maps them to the subscribed SNSSAIs. Furthermore, it determines the set of AMFs serving the UE, or alternatively, the list of candidate AMFs. For the latter task, the NSSF may make query to the NRF (Figures 6.15–6.18).

6.3.10 NWDAF

Network Data Analytics Function (NWDAF) refers to network analytics logical function, which 5G MNO manages. The task of the NWDAF is to provide slice-specific network data analytics to NFs, which are subscribed to it. The data analytics may include e.g. the metrics of load of the network on slide instance level in such a way that the NWDAF does not need to be aware of the actual subscriber of the slice.

Figure 6.15 The interfaces of NRF (which can be divided into home and visited NRF).

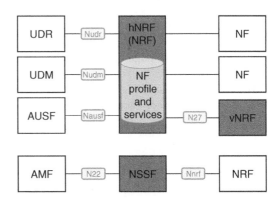

Figure 6.16 The interfaces of NSSF.

Figure 6.17 The interfaces of NWDAF.

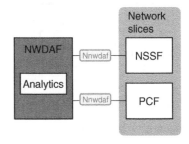

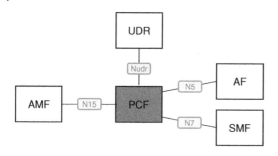

Figure 6.18 The interfaces of PCF.

Please note that in the 3GPP Release 15, the only network functions defined to interface with the NWDAF via the standardized *Nnwdaf* are PCF for policy decision-making and NSSF for slice selection based on the load information. Furthermore, the utilization of the NWDAF for non-slice-specific analytics is not supported in the Release 15.

6.3.11 PCF

The 5G Policy Control Function is an evolution of the LTE PCRF (Policy and Charging Enforcement Function). In 5G, it supports unified policy framework to govern network behavior. It also provides policy rules to control plane functions to enforce them. Furthermore, it can access subscription information relevant for policy decisions in a UDR that resides within the same PLMN. The description of the PCF is in 3GPP TS 23.503.

6.3.12 SEPP

The 5G Security Edge Protection Proxy (SEPP) is a new element for securely interconnecting 5G networks. Figure 6.19 depicts the principle of the utilization of SEPP elements presented in service-based architecture model. The visited network's Security Edge Protection Proxy (vSEPP) refers to the SEPP functionality in visited PLMN while home Security Edge Protection Proxy (hSEPP) refers to the SEPP located at the home PLMN.

The 5G system has designed in such a way that the inter-operator connection via IPX (Internet Protocol Packet eXchange) is protected according to the SBA. IPX has

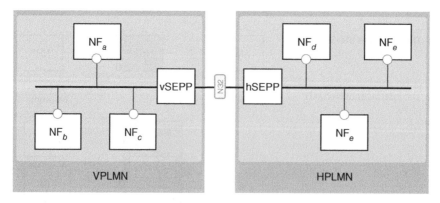

Figure 6.19 The principle of SEPP for interconnecting 3GPP networks.

emerged as a fundamental network backbone for LTE, and it continues being relevant in 5G era, too. Details on IPX in practice can be found in GSMA reference [25].

The SEPP is, in practice, a nontransparent proxy. The role of the SEPP is to be a service relay between Service Producers and Service Consumers, which is a comparable task to direct service interaction. Control plane messages between the SEPP elements may pass through IPX entities.

The SEPP elements are located at the edge of the 5G core network of the MNO. The tasks of the SEPP are to provide confidentiality and integrity protection of the signaling between networks. The SEPP also takes care of the topology hiding of the firewalls according to the defined policies. It can also manage message filtering and policing on inter-PLMN control plane interfaces [21].

The SEPP provides end-to-end application-layer security for the signaling (e.g. security-sensitive IEs, i.e. information elements). The SEPP also takes care of the transport-layer security, referring to hop-by-hop security of adjacent nodes of the IPX network.

3GPP TS 33.501 and TS 29.500 detail the functionality of SEPP, its flows, and the *N32* reference point.

6.3.13 SMF

The 5G SMF replaces together with the 5G UDF the LTE S-GW and P-GW. SMF is defined in 3GPP TS 23.501.

The SMF handles SM-related tasks so it can perform session establishment, modification and releasing. This includes the maintenance of tunnel in UPF–AN node interface. The SMF also performs UE IP address allocation and management, and optionally related authorization. It includes DHCPv4 and DHCPv6 server and client functions.

Furthermore, the SMF performs ARP proxying according to IETF RFC 1027 and/or IPv6 Neighbor Solicitation Proxying according to IETF RFC 4861 for the ethernet PDUs. It selects and controls UP function, makes traffic steering and routing at UPF, terminates interfaces with policy control functions. It includes lawful intercept for SM events and the interface to LI system. It also collects charging data, and controls charging data collection at UPF. It terminates SM-parts of NAS messages, has downlink data notification, includes initiator of AN specific SM information (transported over *N2* via AMF to AN). It determines SSC mode of a session.

Furthermore, SMF includes roaming functionality such as handling of local enforcement for QoS SLA of visited public land mobile network (VPLMN), collection and exposing of charging data (VPLMN), lawful intercept (VPLMN for SM events and interface to LI system), and interaction with external DN for signaling of PDU session authorization/authentication by external DN.

6.3.14 SMSF

The SMSF supports SMS over NAS in 5G infrastructure, managing SMS subscription data and its delivery. It includes SM-RP (Short Message Relay Protocol) and SM-CP (Short Message Control Protocol) with the UE as described in 3GPP TS 24.011 [26].

It includes the procedures necessary to support the SMS between the mobile station (MS) and the mobile switching center (MSC), Serving GPRS Support Node (SGSN),

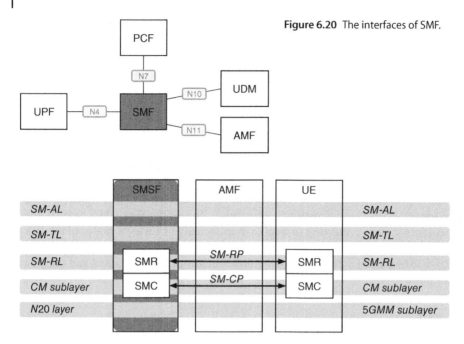

Figure 6.20 The interfaces of SMF.

Figure 6.21 The protocols for the N1 mode used in 5G SMS delivery.

MME, or SMSF, as described in 3GPP TS 23.040 [27]. It also relays the short message between UE and SMS-GMSC, IWMSC, and SMS-router. It collects charging data records (CDRs) on SMS events and performs LI. It interacts with AMF and SMS-GMSC when UE is unavailable for short message delivery by notifying SMS-GMSC which in turn can notify UDM about the unavailability (Figure 6.20).

Figure 6.21 depicts the protocols for the N1 mode used in 5G SMS delivery. In this mode, the Connection Management Sublayer (CM sub) provides services to the Short Message Relay Layer (SM RL). The SM RL provides services to the Short Message Transfer Layer (SM TL) at the MS side while the Short Message Application Layer (SM AL) is the highest protocol layer. The Short Message user information elements are mapped to TCAP (Transaction Capabilities Application Part) or MAP (Mobile Application Part) in the network side.

The MAP is an SS7 protocol providing an application layer for multiple mobile core network nodes to provide services to users. Among other tasks, MAP can be used to deliver short messages within the network. MAP relies on TCAP, and can be transported using the old SS7 protocols, or nowadays, over IP relying on Transport Independent Signaling Connection Control Part (TI-SCCP) or, via SIGTRAN, which is the default in modern mobile networks including 5G.

The protocol between two Short Message Control (SMC) entities is denoted SM CP, while there is SM RP in use between two Short Message Relay (SMR) entities. AMF transfers the SM CP messages between SMSF and the UE.

As depicted in Figure 6.21, there is SM RP between the SMR of SMSF and UE, whereas SM CP is in use between the SMC entities of these same elements. Following the notation of Figure 6.21, MM sub refers to the Mobility Management sublayer, RR sub refers to the Radio Resource Management sublayer, and 5GMM-sub refers to the 5G Mobility Management sublayer.

6.3.15 UDM

As described in 3GPP TS 23.501, the UDM generates the 3GPP AKA Authentication Credentials per user. The UDM resides in the same Home Public Land Mobile Network (HPLMN) with the subscriber it serves. It performs user identification, including storage and management of SUPI per each individual 5G subscriber. Furthermore, it is capable of de-concealing the SUCI, which is the privacy-protected subscription identifier, i.e. the secret version of the SUPI. The UDM also manages access authorizations such as roaming restriction,

UE's Serving NF registration management (RM) (e.g. storing serving AMF for UE, storing serving SMF for UE's PDU session). Furthermore, UDM supports service and session continuity, and provides support to the delivery of mobile terminated short messages. It includes Lawful Intercept Functionality, subscription, and SMS management.

The UDM relies on subscription and authentication data, which may be stored in UDR. This means that multiple UDM elements can provide service to the single user performing various transactions when the UDR is in the same PLMN as the UDM. The UDM can interact also with HSS, although this interface has been left open for implementation.

The 5G UDM is an evolution of HSS and UDR utilized in LTE. It is for 5G data storing and retrieving. As the 5G architecture is based on the separation of data and control planes, e.g. the network status information and other relevant data can be stored in a unified database. The 5G network functions have the right to access selected data of the UDM via standard interfaces to locally store it in a dynamic way.

The distributed database synchronization allows real-time backup procedure between data centers for storing e.g. network status. The unified database simplifies the procedure for network information retrieval functions the benefit being the reduction of signaling overhead on data synchronization. Figure 6.22 clarifies the principle of the UDM in connection with the local data storages.

6.3.16 UDR

The 5G UDR is an evolution of the LTE SDS (Structured Data Storage). As described in 3GPP TS 23.501, the UDM can store and retrieve subscription data from and to UDR. The UDR also takes care of storage and retrieval of policy data by the PCF, storage, and retrieval of structured data for exposure and application data by the NEF.

The UDR resides in the same PLMN as the NF service consumers are. It stores and retrieves data via *Nudr*, which is intra-PLMN interface. In practice, the UDR can be collocated with the unstructured data storage function (UDSF). Figure 6.23 summarizes the interfaces of the UDR.

6.3.17 UDSF

The 5G UDSF is a function comparable with LTE SDSF (Structured Data Storage Network Function). In 5G, the UDSF is left as an optional function for storing and retrieval

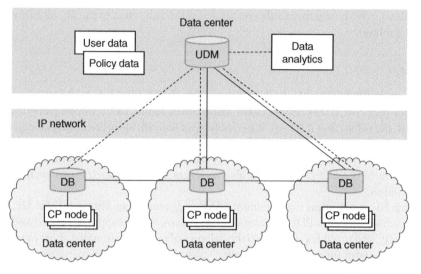

Figure 6.22 A conceptual example of UDM deployment.

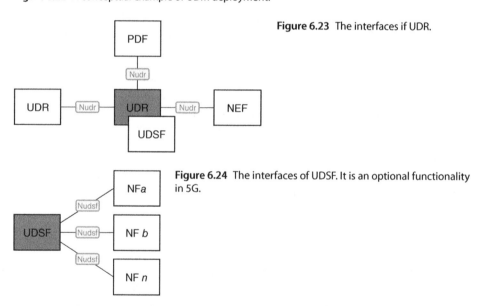

Figure 6.23 The interfaces if UDR.

Figure 6.24 The interfaces of UDSF. It is an optional functionality in 5G.

of information as unstructured data by any NF. Figure 6.24 summarizes the interfaces of UDSF.

6.3.18 UPF

Along with the SMF, the 5G UPF replaces the LTE S-GW and P-GW. The UPF acts as an anchor point for intra- and inter-RAT mobility as appropriate. It also is the external PDU session point of interconnect to the data network and takes care of packet routing and forwarding. It performs packet inspection and takes care of the user plane part of policy rule enforcement and LI. Furthermore, it performs traffic reporting and QoS

Figure 6.25 The interfaces of UPF.

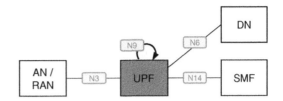

handling for user plane, including data rate enforcement. The UPF verifies uplink traffic by mapping SDF to QoS flow and makes transport-level packet marking. It buffers downlink packets and triggers downlink data notifications, and it also sends and forwards end markers to the source NG-RAN node.

The UPF can perform ARP proxying according to the IETF RFC 1027 and IPv6 Neighbor Solicitation Proxying according to IETF RFC 4861. Figure 6.25 summarizes the interfaces of UPF.

6.4 5G Functionalities Implemented in 5G Core

6.4.1 Network Slicing

The 5G era has advanced and new business models. There are much more possibilities to cooperate between MNOs and third parties in such a way that some external organization gets right to utilize a dedicated part of the network. The respective use cases may relate to public safety and CriC to ensure people's safety. This type of dedicated resource utilization can be done via the network slicing, which is one of the most essential concepts in the transition toward 5G core network.

The network slicing provides the possibility of isolating network resources for certain services. The 5G system can execute resource provisioning so that certain set of the logical network nodes are assigned to the third party. Figure 6.26 depicts an example of such principle by dedicating the UPF, SMF, and AMF resources within a set of isolated network slices for each cooperating organization relying on those slices.

Also, the network slice concept allows the differentiation of the level of the provisioned security per service type, which are taken care of by different slices. This is a remarkable difference with the previous mobile generations because before, the applied security would be the same for all the users (devices), regardless of the service type. Although this has not been a limitation up to now, it is not optimal either, as the level of the applied security in networks prior to the 5G may be unnecessarily strong for certain cases such as simple IoT communications. Also, the uniformly applied security may be considered weak for some other service types such as the CriC.

The 5G network slicing solves these problematics of ideal security level by distinguishing the level per use case. Furthermore, the resource utilization is optimal as the control plane of the 5G can be managed by the cloud, ensuring flexible and interrupt-persistent functioning via geo-redundant configuration. The user plane nodes, in turn, can work on services within dedicated slices. The slice can be instantiated, updated, and deleted according to the NFV concept, which makes it possible to differentiate the QoS in a fluent way. Figure 6.27 presents an example of such implementation.

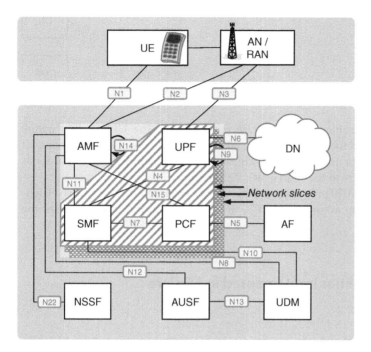

Figure 6.26 The principle of the network slicing in core network deployment.

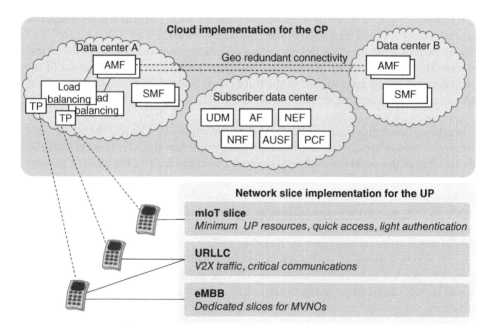

Figure 6.27 Example of the network slice set and cloud implementation.

6.4.2 SDN

The SDN refers to network architecture that has been designed to minimize the hardware constraints by abstraction of the low-level functions. Instead, the SDN makes it possible to execute these functions in a software-based, centralized control plane via API. The benefit of this approach is that the network services are agnostic to the underlaying hardware, and they can be offered and utilized regardless of the connected hardware elements.

Specifically, for 5G, the SDN concept can be used as a framework that makes it possible to provide 5G to function in the control plane, thus optimizing the data transmission. As an example, instead of using dedicated network elements such as policy control functions, which is a standalone component in earlier networks, 5G enables the functionality of such component to perform its tasks on a common hardware shared amongst many other network functions.

Some of the benefits of the SDN concept are the optimized bandwidth and enhanced latency performance. Furthermore, the rerouting of the data flows takes place in practically real time via the SDN, which considerably enhances the prevention of network outages and thus contributes to the development of high-availability services that are desired e.g. in the CriC of 5G.

6.4.3 NFV

Together with the SDN concept, the NFV has an important role in providing optimal 5G performance. The principal task of the NFV is to decouple the software from the hardware via the deployment of VM [28]. These VMs perform a multitude of network functions such as encryption of the communications. The benefit of VM concept is the highly dynamic characteristics, so whenever needed, the VM is created for the respective function automatically. This results in cost savings, as the operator of such network functions does not need to invest in a standalone element, which typically is based on proprietary hardware architectures. Yet another benefit of the virtualization of the network functions is the possibility of deploying them much faster than the dedicated HW/SW elements.

As for 5G, the NFV enables the network slicing concept, which is an elemental part of the system. Network slicing enables the use of multitude of virtual networks on top of a physical infrastructure, and it also provides means for dividing a physical network into multiple virtual networks, which can serve various RANs. Yet another benefit of the NFV is the possibility to easily scale the functions based on different criteria for optimal performance, e.g. as for the cost or energy consumption.

6.4.4 Optimization of the Core and Transport

The methods already mentioned are an integral part of optimally performing 5G. There are also other methods that can be applied to the 5G core and transport networks to further enhance the capacity and service availability. Such techniques include programmable transport, transport-aware radio, and dynamic load balancing.

5G network performance can benefit for the *programmable transport,* as the transport resources can be allocated dynamically upon the need and current capacity consumption. This concept can be applied in a locally variable environments for, e.g. managing energy consumption dynamically. Some use cases, as mentioned in [29], are network sharing for offering flexible service provisioning and optimization of the utilized resources via dynamic resource allocation.

The second optimization method mentioned in [29] is the *transport-aware radio*. It refers to the enhanced functionality in X2 interface, i.e. between base station elements. Traditionally, the handover can be performed fluently via the X2 interface, but if the respective transport network is experiencing congestion, there will be service degradation in such intention. In fact, the handover in such a situation would make the situation worse. The transport-aware radio functionality may thus conclude that the overall performance is better without X2 signaling while the congestion takes effect, and advice the network accordingly.

Third, optimization method is the dynamic load balancing as described in [29]. This method is beneficial e.g. in nonpredictable weather conditions that might negatively affect the radio link, lowering the planned data rate due to the attenuation of rain. The offered load can be observed via service orchestration, which analyzes the ongoing level, and the potential impacts on the SLA. As the exceeding of the agreed SLA levels typically leads into additional expenses via noncompliance fees, the service orchestration can have embedded intelligence to evaluate the cost-impact. In the negative case, the service orchestration can analyze the alternative paths and their respective cost impacts, to decide dynamically the most cost-efficient option, and to execute an order to change the data flow path accordingly.

6.4.5 Cloud RAN

5G will bring along with new RAN architectures. In addition to the established concept of distributed RAN mode, 5G can be deployed also by centralized and cloud RAN, and there will thus be new fronthaul interfaces defined. Figure 6.28 summarizes these models [30].

The cloud concept can be applied in the core network of 5G system as well as in the radio network [31]. The cloud concept applied so widely, supported by the SDN and NFV concepts, further contributes to the flexible compliance of the demanding performance requirements. The combination of SDN and NFV effectively makes the network infrastructure cloud-based as for the access, transport, and core network.

The benefit of the cloud concept is the fluent support of highly diversified services. It also serves as a base for the network slicing and dynamic deployment of services.

The components of the cloud RAN are the physical RAN/AN sites (including, e.g. 3GPP radio-based stations and Wi-Fi-based stations), and mobile cloud engines (MCEs), as described in [32]. The task of the MCE is to coordinate services of real-time and non-real-time resources of various access technologies.

The cloud RAN connects to the service-oriented core network, i.e. core cloud as depicted in Figure 6.29. The core network has multiple functions such as dynamic policy control and storing of data in the unified database.

Furthermore, there is a control plane with a set of components for performing the selected network functions, as well as programmable user plane fulfilling the variety of service requirements. This concept provides the possibility of orchestrating the network functions for the selection of relevant control and user plane functions.

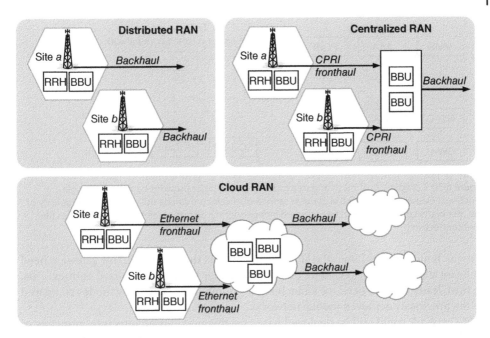

Figure 6.28 The models for 5G RAN architectures.

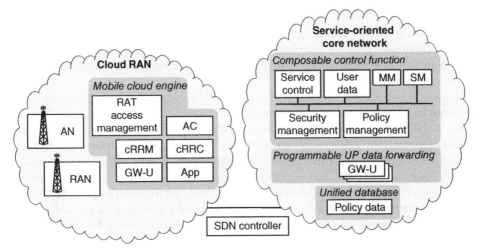

Figure 6.29 The principle of the cloud RAN.

As can be seen in Figure 6.29, the SDN controller actuates in the transmission path between the cloud RAN and core network. The transmission infrastructure also contains a set of data forwarding paths that depend on the service requirements.

6.4.6 User and Data Plane Separation

5G will optimize the user and control plane functions. The network gateways prior to the 5G era managed both user and control plane functions. Nevertheless, 5G services

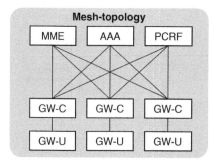
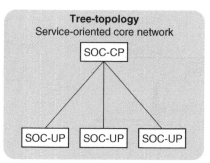

Figure 6.30 Comparison of mesh and tree topologies. The latter provides benefits as a single functional element for the control plane of service-oriented core can be utilized while the capacity of the user plane components can be increased independently; also, the tree-topology simplifies the core connectivity.

are much more critical than the latency. This means that the respective gateways need to be located toward data centers that may be either local or centralized entities. This, in turn, leads into the need for a much higher number of the gateway nodes compared to the previously deployed mobile networks.

5G network could still be deployed based on this previous architectural principle of mesh network topology, but the drawback of it is the increased complexity as for the configuration of the gateway services; the increased number of distributed gateways also increases the number of the respective links. This, in turn, leads to an unfavorable situation, as the respective signaling may be significant in the path from the base station through distributed gateway and network's control plane. As a result, the link and handover signaling in such topology is significant and negatively affects both capital and operating expenditure of the MNO.

Figure 6.30 also depicts the more favorable tree-topology that provides optimal performance for the 5G signaling, separating the user and control planes.

As can be seen in Figure 6.30, the tree topology of the service-oriented 5G core network separates the gateway user and control planes so that they perform and can be, e.g. upgraded independently from each other, making the scaling of the network much more efficient and flexible. This distributed approach provides cost-efficient gateway deployment. Furthermore, it reduces the signaling load and the overall signaling routes. Other benefits of the separation of the user and control planes include the utilization of centralized control logic functions and efficient network slicing for a variety of applications [32].

In practice, the separation of the user and control plane requires functions supporting the division of the control logics. Furthermore, there is a need for object-oriented interfaces for the user plane to make it programmable, and finally, the orchestration of the service objects takes place in the control plane.

6.5 Transport Network

The role of the transport network is to interconnect the radio access and core network. To comply with strict performance requirements for the 5G networks, the optimal deployment can be done by relying on increased intelligence, flexibility, and automation in the transport network [33].

As the RAN and core network develop, the transport network also needs to follow the evolution in order not to create a bottleneck, as for the QoS level and performance. This requires a higher grade of flexibility, simple service configurations, and new operations model support and cross-domain orchestration. The dynamic and virtualized RAN requires that the transport connectivity is more dynamic than in previous generations, which requires inevitably more automatization to manage the transport.

As an example of the commercial productization, Ericsson has developed SDN, accompanied by an intelligent application referred to as transport intelligent function (TIF), in an effort to optimize the 5G transport network design. In this solution, the 5G transport network is based on a self-contained infrastructure underlay with an SDN-controlled overlay for a variety of RAN and user services. The distributed control plane in the underlay maintains the basic infrastructure and handles redundancy and quick restoration if the network fails. The service- and characteristics-aware overlay is handled by the SDN controller with the TIF application, and this creates a dynamically controlled and orchestrated transport network that requires minimum manual interaction [33]. The respective automation framework of the solution is based on the Open Network Automation Platform (ONAP).

6.5.1 Conceptual Examples

6.5.1.1 Cross-Domain Communication
Figure 6.31 depicts a transport solution presented in [33]. In this figure, the TIF is a cross-domain function located between radio access and transport networks.

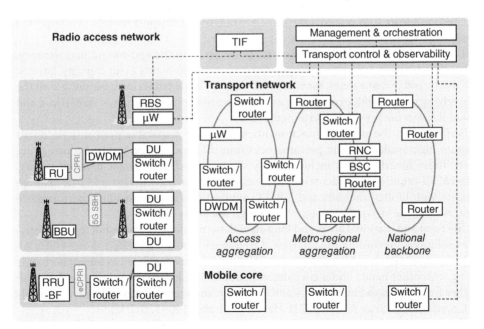

Figure 6.31 Conceptual example of 5G transport solution as presented by Ericsson [WP].

Interpreting further the Ref. [33], the TIF receives RAN connectivity requests via cross-domain communication bus. The TIF processes the requests with preset policies and performance-aware topology data within the control architecture, as depicted in Figure 6.31. Ref. [33] continues with a description of a transport automation framework. It is designed to support the following stages:

- *Automatic installation* of transport network infrastructure by (i) establishing configuration with O&M connectivity to the controller; (ii) registering with control and management functions.
- *Automatic configuring* of transport connectivity service for the desired RAN interfaces interacting with the RAN as connectivity requirements are considered.
- *Automatic optimization* of transport resources during the run-time phase to achieve overall optimal quality of experience (QoE) for consumers interacting with the RAN domain to achieve overall optimization. While coordinated RAN transport automation is useful for auto-integration, its benefits are much more significant for auto-configuration and auto-optimization.

Continuing the practical examples from the industry for the 5G transmission, Nokia has analyzed the microwave solutions.

6.5.1.2 Microwave Transport Network

The hardened requirements for the 5G transport are a result of enhanced broadband, massive machine-to-machine communications, and critical machine communication modes of 5G. Each of them has special needs and affects the whole transmission chain. Furthermore, the technical realization of the transmission may take many different forms, from fixed fiber optics ideal for dense city areas and suburban areas while radio links may be the most adequate solution for many rural and remote locations. The example in Figure 6.28 presents the principle of typical transport models as interpreted from Ref. [30].

The transport network must not create bottlenecks in the end-to-end performance. It has thus typically very demanding performance requirements as for capacity, latency, reliability, and in many cases, for energy efficiency. The demands increase along with the 5G, relying more on edge computing as well as on to the introduction of centralized and cloud RAN to the previous distributed RAN.

Furthermore, the new 5G functionalities and their respective performance needs require high availability and performance from the modernized transport infrastructure. These functionalities include highly dynamic and performant network slicing. The eMBB requires high increased data rates and capacity, whereas mIoT needs high density with limited mobility and URLLC applications require excellent performance figures from the viewpoint of latency and high reliability.

The rural and marginal areas and long-haul data transfer all require microwave transport solutions. According to [30], typical frequency bands are located above 4 GHz up to V-band (60 GHz) and E-band (70–80 GHz), which are considered as mm-wave bands along with other bands under consideration above 42 GHz. In the future, there may be higher frequencies defined in the standardization and included into the practical product portfolios, such as W-band (90 GHz) and D-band (140–170 GHz).

The transport frequency channels vary in bandwidth. In 5G, the practical maximum bandwidth may be up to 2 GHz in the highest-frequency bands.

The bandwidth contributes directly to the achievable maximum bit rate. As an example, the typical bit rates in unlicensed V-band of 60 and 0.5 GHz bandwidth may be up to 1 Gb s^{-1}, whereas W-band of 90 and 2 GHz bandwidth provide data rates up to about 10 Gb s^{-1}. Logically, the higher the frequency band, the less coverage it provides according to the radio propagation characteristics per geographical topology. It may be that the 90 GHz band provides only a few hundreds of meters coverage for the highest bit rates, while traditionally used sub-6 GHz radio links achieve tens of kilometers, up to 100–150 km distances, by applying highly directional link antennas.

6.6 Protocols and Interfaces

The 5G protocol layer consists of the variants present in each reference point of the new architecture.

6.6.1 RAN

Table 6.2 summarizes the 5G RAN interfaces.

The protocol layers of *NG* and *Xn* for the control and user plane are depicted in Figure 6.32. The *Xn* application protocol (AP) serves in the role of application layer signaling protocol. The transport layer protocol for the control plane of *Xn* in Stream Control Transmission Protocol (SCTP) while the user plane relies on the GDP-U, respectively.

NG-AP refers to NG AP. It is defined in 3GPP TS 38.423 and TS 38.424 (*Xn* data transport). SCTP is a reliable transport layer protocol defined by the Internet Engineering Task Force (IETF). It operates on top of a connectionless packet network such as Internet protocol (IP). IP refers to Internet protocol as defined by IETF. The data link layer is for the data transport of the NR. It is defined in 3GPP TS 38.414. The next-generation

Table 6.2 5G NR-RAN interfaces as per 3GPP technical specifications.

Interface	Description	Source
NG	*NG* refers to *New Radio,* which is renewed radio interface between the UE and gNB. It is based on OFDM both in downlink as well as in uplink.	TS 38.410...38.414, *NG* interface general aspects and principles
Xn	Interface is between NG-RAN nodes gNB and ng-eNB.	TS 38.420...38.424, *Xn* interface general aspects and principles
E1	E1 is point-to-point interface between gNB-CU-CP and a gNB-CU-UP.	TS 38.460...38.463
F1	Interface between gNB-CU and gNB-DU.	TS 38.470...38.475, *F1* interface general aspects and principles
F2	Interface between lower and upper parts of the 5G NR physical layer; includes F2-C and F2-U.	

Source: 3GPP TS 38.401.

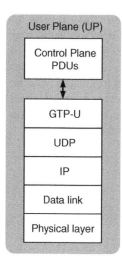 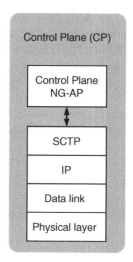

Figure 6.32 The protocol stacks for *NG* and *Xn* interfaces.

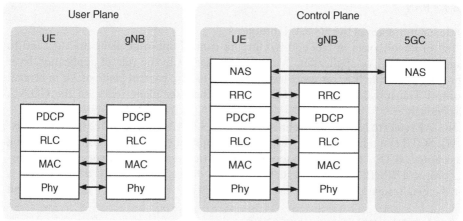

Figure 6.33 The high-level 5G user and control plane protocols.

physical layer, i.e. the physical aspects of the NR, is defined in 3GPP TS 38.411 and 38.421 TS (*Xn* layer 1).

6.6.2 Layers

The 5G protocol stack consists of layers 1, 2, and 3. Layer 1 refers to the physical layer. Layer 2 includes MAC, radio link control (RLC) and packet data convergence protocol (PDCP), while layer 3 refers to the radio resource control (RRC) layer [34]. Figure 6.33 summarizes the high-level 5G protocol stacks on user and control planes.

Figure 6.34 presents the frame structure of the RLC and packet data convergence protocol (PDCP).

6.6.3 Layer 1

The 5G layer 1, i.e. the physical layer, includes the following main functionalities:

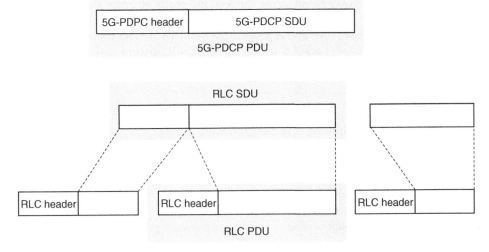

Figure 6.34 The frame structure of RLC and PDCP.

- Transport channels' error detection, Forward Error Coding (FEC) and decoding
- Soft-combining of Hybrid Automatic Repeat Request (HARQ)
- Coded transport and physical channels' rate matching and mapping
- Physical channels' power weighting, modulation, and demodulation
- Synchronization in frequency and time domains
- RF processing, radio measurements, and reporting of radio characteristics
- multiple-in, multiple-out (MIMO) antenna processing, antenna beamforming, and transmit diversity management

6.6.4 Layer 2

Layer 2 refers to 5G MAC, RLC, and PDPC functionalities. For MAC, these include:

- Antenna beam management
- Random access procedures
- Logical and transport channel mapping
- MAC Service Data Unit (SDU) concatenation from logical channel into transport block (TB)
- 5G MAC SDU multiplexing and demultiplexing
- Reporting of scheduling information
- Hybrid automatic repeat and request (HARQ)-based error correction
- UE's logical channel priority management, and UE priority management by dynamic scheduling
- Transport format selection
- Padding

For the RLC, the functions of layer 2 include:

- Upper-layer PDU transfer
- 5G-RLC reestablishment

- AM data transfer's ARQ-based error correction, protocol error detection, and re-segmentation
- Unacknowledged Mode (UM) and Acknowledge Mode (AM) data transfer's 5G RLC PDU reordering, duplicate packet detection, 5G-RLC SDU discard, and segmentation

For the PDPC, the functions of layer 2 include:

- User data transfer
- Upper-layer PDU in-sequence delivery at the 5G-PDCP reestablishment procedure, and lower-layer SDU duplicate detection, for 5G-RLC AM
- 5G-PDCP SDU retransmission in the connected 5G-RLC mobility mode
- Ciphering and deciphering based on mandatory Advanced Encryption Standard (AES)
- User plane ciphering and integrity protection
- Uplink SDU discard based on timer value
- Control plane data transfer

6.6.5   Layer 3

The 5G layer 3, i.e. the RRC layer, includes the following main functionalities:

- System information broadcasting for NAS and AS
- RRC connection establishment, maintenance, and release
- Key management and other security procedures
- Point-to-point (PTP) radio bearer establishment, configuration, maintenance, and release
- Mobility functions
- Radio interface measurements and reporting from the UE
- Message transfer between NAS and UE

6.6.6   The Split-Architecture of RAN

Figure 6.35 depicts the split-architecture of gNB. The functions of gNB can be divided by applying centralized units (CUs) for user plane (UP) and control plane (CP), while the UP and CP are common for the distributed unit (DU) and remote radio head (RRH) of the gNB.

6.6.7   Core Network Protocols

This section summarizes the 5G-specific network reference points and protocols between the network functions. The list of key reference points is presented in Table 6.3.

Some of the key specifications to detail the 5G core network and the respective protocols used in the 5G core are:

- 3GPP TS 23.231, IP-I based circuit-switched core network; Stage 2
- 3GPP TS 23.501, System Architecture for the 5G system, which includes network functions and procedures
- 3GPP TS 23.502, Procedures for the 5G system

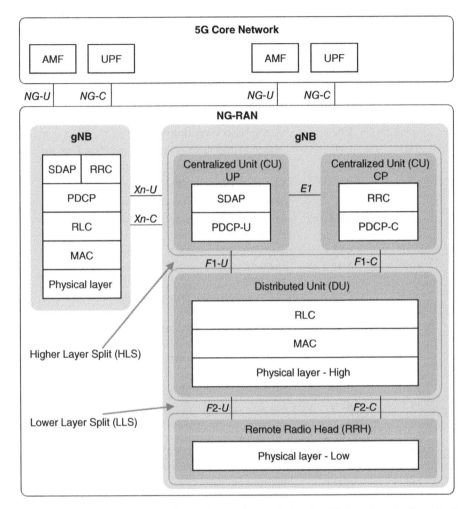

Figure 6.35 The functional split of user plane and control plane in gNB based on distributed unit.

- Verizon 5G Technical Forum TS V5G.300 [35]
- Verizon 5G Technical Forum TS V5G.201 [36]

The 3GPP TS 23.501 details the protocol stacks of 5G for the interfaces. The following sections summarize the control and user plane protocols of the key components based on these information sources.

6.6.8 Control Plane Protocols

6.6.8.1 Access Stratum for *N2*
The access stratum (AS) control plane protocols between 5G access network and 5G core network take place on *N2* interface as defined in 3GPP TS 38.413. The *N2* interface includes the respective interface management procedures such as the configuration and resetting the interface. The tasks related to a specific UE include procedures for NAS

Table 6.3 Reference points of 5G.

Interface	Element *a*	Element *b*	Description
NG1	UE	AMF	User Equipment – Access and Mobility Management Function
NG2	AN/RAN	AMF	(R)AN – Access and Mobility Management Function.
NG3	AN/RAN	UPF	(R)AN – User Plane Function
NG4	SMF	UPF	Session Management Function – User Plane Function
NG5	PCF	AF	Policy Control Function – Application Function
NG6	UPF	DN	User Plane Function – Data Network
NG7	SMF	PCF	Session Management Function – Policy Control Function
NG8	UDM	AMF	Unified Data Management – Access and Mobility Management Function
NG9	UPF	UPF	Two User Plane Functions
NG10	UDM	SMF	Unified Data Management – Session Management Function
NG11	AMF	SMF	Access and Mobility Management Function – Session Management Function
NG12	AMF	AUSF	Access and Mobility Management Function – Authentication Server Function
NG13	UDM	AUSF	Unified Data Management – Authentication Server Function
NG14	AMF	AMF	Access and Mobility Management Functions
NG15	PCF	AMF	a) *Non-roaming scenario.* Policy Control Function – Access and Mobility Management Function b) *Roaming scenario.* V-PCF – AMF
N27	hNRF	vNRF	NRF in the home PLMN is known as the hNRF, which is referenced by the vNRF via the *N27* interface
N32	hSEPP	vSEPP	Roaming interface
NWu	UE	AMF	N3IWF relays the IPsec between UE and AMF

transport, UE context management, procedures with resources for PDU sessions and handover management.

The control plane interface between the 5G access network and the 5G core supports multiple 5G access network technologies such as 3GPP RAN for trusted access and N3IWFs for untrusted access to the 5G core relying on NGAP protocol for all the access types on the *N2* interface. The NGAP terminates at AMF for each UE and their accesses for any number of their PDU sessions. The NGAP can contain instructions for AMF to relay the data between the 5G-AN and the SMF. This is referred to as *N2* SM information, and it is detailed in 3GPP TS 38.413 and TS 23.502.

Figure 6.36 summarizes the protocol stack applied in the *N2* between the access network and AMF.

As can be seen in Figure 6.36, the top-level protocol of *N2* is the NG Application Protocol (NG-AP), which is detailed in 3GPP TS 38.413. The underlaying SCTP ensures the signaling message delivery, and it is detailed in IETF RFC 4960.

Figure 6.36 The control plane protocols of *N2* interface.

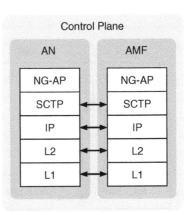

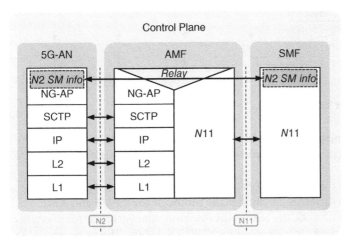

Figure 6.37 The control plane protocols for the interfaces *N2* and *N11*.

Figure 6.37 summarizes the control plane protocols between the AN and SMF when AMF is functioning as a relay in between.

As can be seen in Figure 6.37, the top-level NG-AP protocol contains *N2* SM information field. It refers to a subset of the NG-AP information, which the AMF relays transparently between the AN and SMF.

6.6.8.2 Non-Access Stratum for UE and 5G Core Network

When the UE connects to the 5G core network, *N1* NAS signaling takes place for each access. The termination point for the *N1* is at AMF.

The *N1* signaling is used for RM, Connection Management (CM), and SM of the UE. The *N1* NAS protocol contains NAS-MM (Mobility Management) and NAS-SM (Session Management).

The tasks of the NAS protocols related to N1-MM include SM signaling, short message delivery, UE policy, and Location-Based Services (LBS). Among other tasks, the NAS-MM supports NAS procedures terminate at AMF, such as handling of RM and CM state machines and procedures with the UE. The AMF supports the related capabilities of *N1* signaling during the RM/CM procedures, routing to other network function

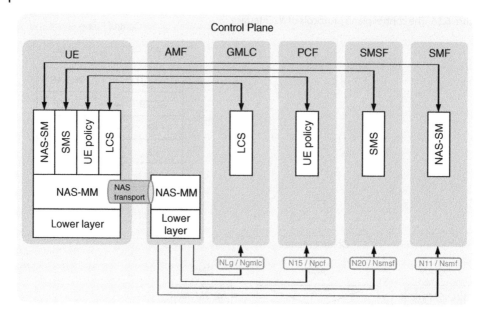

Figure 6.38 The NAS transport for selected uses.

or processing locally the messages. The AMF also provides a secure NAS signaling connection by applying integrity protection and ciphering of the messages between the UE and the AMF.

Figure 6.38 summarizes the NAS transport for selected cases.

Figure 6.39 summarizes the NAS signaling plane protocols for the UE and AMF.

As can be seen from Figure 6.39, the NAS-MM (Mobility Management) functionality is on top of the protocol stack. It is detailed in 3GPP TS 24.501. The respective NAS protocol supports registration and CM as well as user plane connection activation and

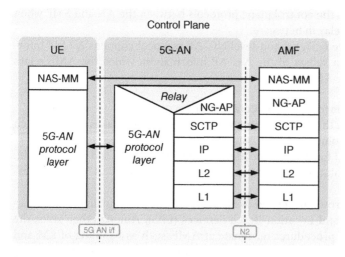

Figure 6.39 The control plane protocols between the UE and AMF.

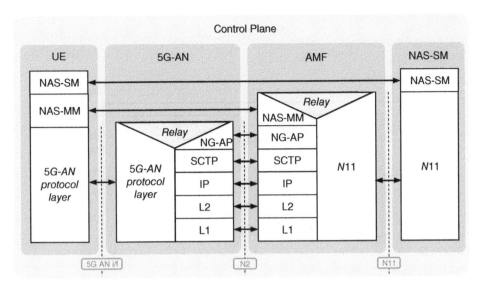

Figure 6.40 Control plane protocol stack for the communications between UE and SMF.

deactivation. The NAS-MM also manages the NAS signaling ciphering and integrity protection.

The radio interface protocol stack between UE and 3GPP 5G-gNB or previous generation's evolved Node B (eNB) is detailed in 3GPP TS 36.300 and TS 38.600 and is also summarized in the Radio Network chapter of this book. For the non-3GPP access, the 5G-AN protocol layer reflects the respective system.

6.6.8.3 UE-SMF

The NAS-SM is applied to the SM between the UE and SMF without AMF interactions. The SM message includes the session ID of each PDU. As can be seen in Figure 6.40, the NAS-SM functionality is in the top of the respective protocol stack. The main tasks of the SM functionality's NAS protocol is to perform user plane PDU session establishment, modification and release (Figure 6.41). The 5G NAS protocol is detailed in 3GPP TS 24.501.

6.6.8.4 Control Plane Protocols for NFs

The control plane protocol stacks for the service-based interfaces are detailed in 3GPP TS 29.500, and the control plane protocols in *N4* interface (SMF – UPF) are descried in 3GPP TS 29.502.

The control plane protocols for the 5G network functions depend the scenario from which Figure 6.42 depicts control plane for the scenario prior to the signaling IPsec SA establishment between UE and N3IWF, Figure 6.43 depicts the protocol stack of control plane after the signaling IPsec SA is established, and finally Figure 6.44 depicts the scenario of Control Plane for establishment of user-plane via N3IWF. In these cases, the User Datagram Protocol (UDP) can be applied for the interface of UE – N3IWF with the aim of providing NAT traversal for IKEv2 and IPsec traffic. This type of signaling is described in 3GPP TS 23.502 (Figure 6.41).

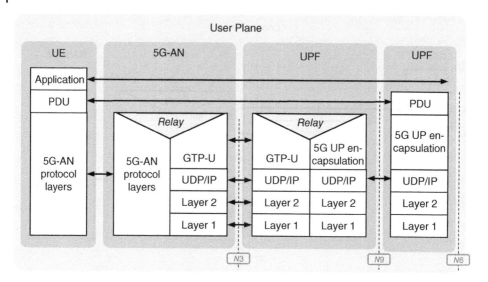

Figure 6.41 The 5G's user plane PDU session protocol structure. The last UPF refers to a non-PDU session anchor and is defined as an optional.

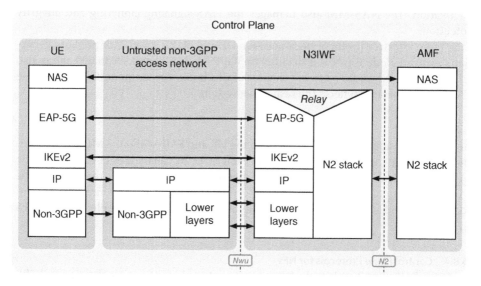

Figure 6.42 Control plane for the scenario prior to the signaling IPsec SA establishment between UE and N3IWF.

6.6.9 User Plane Protocols

6.6.9.1 PDU Session

The user plane protocols via the 3GPP access include scenarios for PDU layer, general packet radio service (GPRS) tunneling and 5G encapsulation.

The PDU layer scenario refers to the PDU transported between the UE and the data network by relying on PDU session. The session can be of IPv4, IPv6, or a combined IPv4V6. It can also be of ethernet, which transports the respective ethernet frames.

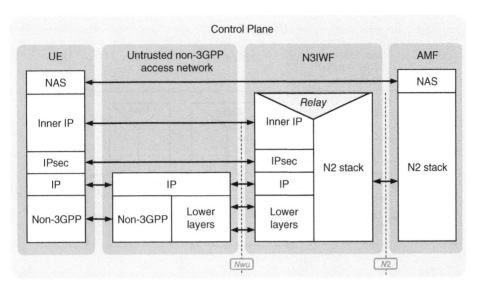

Figure 6.43 The protocol stack of control plane after the signaling IPsec SA is established between the UE and N3IWF. The IPsec between UE and N3IWF refers to the tunnel mode.

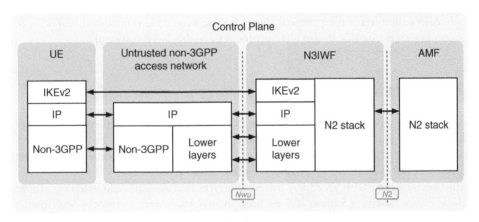

Figure 6.44 Control plane for establishment of user-plane via N3IWF.

The GPRS Tunneling Protocol in User Plane is referred to as GTP U. It is capable of multiplexing PDU session traffic by tunneling *N3* user data transported in backbone network between the 5G access network and the UPF. The GTP encapsulates the PDUs of each PDU session and makes the marking for QoS flows.

The 5G encapsulation layer multiplexes and encapsulates *N9* PDU session traffic of PDU sessions between UPF and 5G core network. It also makes marking for QoS flows. The *N9* can refer to both inter-PLMN and intra-PLMN scenarios.

The use of 5G-AN protocols depends on the access scenario. The 3GPP TS 38.401 defines the 3GPP NG-RAN scenario, and the 3GPP TS 36.300 and TS 38.300 define the radio protocols for UE–eNB and UE–gNB. The role of the N3IWF is to connect 5G core network with untrusted non-3GPP access network.

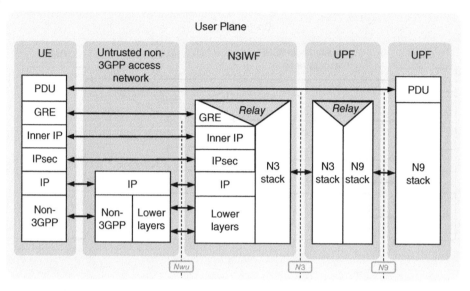

Figure 6.45 User plane for N3IWF access.

6.6.9.2 User Plane for Non-3GPP Access
Figure 6.45 summarizes the protocol stacks for the access via nontrusted N3IWFs. In this scenario, it is possible to use UDP protocol to enable NAT traversal.

6.7 5G Core Network Services

There are many services defined in previous 3GPP releases, which keep evolving and are valid also in 5G era, in a parallel fashion. Some of the key services are the many-fold IoT modes, including LTE-M and NB-IoT, as well as evolved MBMS (Multimedia Broadcast Multicast Service) and CBRS (Citizens Broadband Radio Service).

The specification 3GPP TS 23.501 lists key services that are still valid in 5G era. These include Public Warning System (PWS), SMS over NAS, Internet protocol Multimedia Subsystem (IMS), Emergency services, multimedia priority services, eCall, and emergency services fallback.

For the IoT prior to the mIoT class of Release 16, the key requirements can be found in ETSI SCP TS 103 465. For the vehicle-to-vehicle (V2V) and vehicle-to-everything (V2X) communications, the key specification for additional information is 3GPP TS 23.285 containing architecture enhancements for V2X services.

For the enhanced MBMS, the key specification is 3GPP 23.246 MBMS. It is defined for 3G and 4G, but it also continues being useful base for respective services in 5G era, as has been reasoned in [37]. The benefit of the eMBMS is the delivery of the same content to all the users located within its service area without needing to constantly acknowledge the reception of files, multimedia contents, and other multi-casted or broadcasted data. It thus optimizes the radio spectrum for the highest demand video-type of transmissions, unlike in unicast mode. It can also be expected to serve the massive amount of IoT devices foreseen soon, e.g. for the SW updates and other common contents delivery over the air (OTA). It can be used for consumer devices such as smart devices for operation system updates and other common OTA-deliveries.

References

1 3GPP, "TR 22.891; Feasibility study on new services and market technology enablers, Release 14," 3GPP, 2016.

2 R. Miller, "2018 With 5G on the Horizon, Telcos Embrace Open Compute," Data Center Frontier. https://datacenterfrontier.com/with-5g-on-the-horizon-telcos-embrace-open-compute. [Accessed 21 June 2018].

3 R. S. Raynovich, "2018 How IoT & 5G will change the data center," Futuriom. http://www.futuriom.com/articles/news/how-5g-iot-will-change-the-data-center/2018/02. [Accessed 21 June 2018].

4 M. Carugi, "2018 Key features and requirements of 5G/IMT-2020 networks," ITU. https://www.itu.int/en/ITU-D/Regional-Presence/ArabStates/Documents/events/2018/RDF/Workshop%20Presentations/Session1/5G-%20IMT2020-presentation-Marco-Carugi-final-reduced.pdf. [Accessed 21 June 2018].

5 Google, "2018 Google Compute Engine Service Level Agreement (SLA)," Google. [Accessed 21 June 2018].

6 Brown, G. (2017). Contextual service assurance in 5G: New requirements, new opportunities. *Heavy Reading.*

7 GSMA (2017). *An Introduction to Network Slicing.* GSM Association.

8 Nokia, "Cloud Mobile Gateway." https://networks.nokia.com/products/mobile-gateway. [Accessed 21 June 2018].

9 Ericsson, "2018 Ericsson Virtual Evolved Packet Core," Ericsson. https://www.ericsson.com/ourportfolio/core-network/virtual-evolved-packet-core?nav=fgb_101_256. [Accessed 21 June 2018].

10 5G-MoNArch Project No. 761445, "2018 Documentation of Requirements and KPIs and Definition of Suitable Evaluation Criteria," 5GPPP, June 8.

11 5G-MoNArch Project No. 761445, "2017 Initial resilience and security analysis," 5GPPP, September 30.

12 Henning, S., Cinzia, S., and Seppo, H. (2012). *LTE Self-Organising Networks (SON).* Hoboken, NJ: Wiley.

13 Sungard, "2018 Data and disaster recovery centers," Sungard. https://www.sungardas.com/en/services/data-and-recovery-centers. Accessed 21 June 2018.

14 Gravic Shadowbase, "2018 Disaster Recovery and Active/Passive Replication Systems," Gravic Shadowbase. https://shadowbasesoftware.com/solutions/business-continuity/disaster-recovery. [Accessed 21 June 2018].

15 Juniper, "2018 Example: Configuring an Active/Active Layer 3 Cluster Deployment." Juniper. https://www.juniper.net/documentation/en_US/release-independent/nce/topics/example/chassis-cluster-srx-active-active-configuring.html. Accessed 21 June 2018.

16 Giesecke+Devrient, "Manage your SIM cards – securely over-the-air," Giesecke+Devrient. https://www.gi-de.com/mobile-security/industries/connectivity-service-providers/ota-solutions. [Accessed 31 July 2018].

17 Giesecke+Devrient, "2018 Manage your SIM cards – securely over-the-air," Giesecke+Devrient. https://www.gi-de.com/en/us/mobile-security/industries/connectivity-service-providers/ota-solutions. [Accessed 21 June 2018].

18 ITU-T, "2017 Cloud computing – Functional architecture of Network as a Service (Recommendation Y.3515)," ITU-T, July.

19 GSMA, "2018 Security Accreditation Scheme," GSMA. https://www.gsma.com/aboutus/leadership/committees-and-groups/working-groups/fraud-security-group/security-accreditation-scheme. [Accessed 21 June 2018].

20 "What Are the Best Data Center Certification Standards?," Colocation America. https://www.colocationamerica.com/data-center-certifications. [Accessed 21 June 2018].

21 3GPP, "TS 23.501; System Architecture for the 5G System; Release 15," 3GPP, 2018.

22 P. Eronen, "2006 IKEv2 Mobility and Multihoming Protocol (MOBIKE)," IETF. https://www.ietf.org/rfc/rfc4555.txt. Accessed 23 June 2018.

23 Peter Schmitt (Huawei) "2017 Control and User Plane Separation of EPC nodes (CUPS)." 3GPP. http://www.3gpp.org/cups. Accessed 23 June 23 2018.

24 K. Rabie, "2017 Core Network Evolution (RCAF, PFDF, & TSSF) - 3GPP REST Interfaces," LinkedIn. https://www.linkedin.com/pulse/core-network-evolution-rcaf-pfdf-tssf-3gpp-rest-interfaces-rabie. Accessed 23 June 2018.

25 GSMA, "IR.34 – Guidelines for IPX Provider networks; Version 9.1," GSMA, 2013.

26 3GPP, "TS 24.011; Point-to-Point (PP) Short Message Service (SMS), Release 15," 3GPP, 2018.

27 3GPP, "TS 23.040; Technical realization of the Short Message Service (SMS), Release 15, V15.1.0," 3GPP, 2018.

28 Cranford, N. (2017). The role of NFV and SDN in 5G,. *RCR Wireless News*, December 4. https://www.rcrwireless.com/20171204/fundamentals/the-role-of-nfv-and-sdn-in-5g-tag27-tag99. [Accessed 31 July 2018].

29 Parssons, G. (2014). *5G, SDN and MBH*. Ericsson.

30 Bougioukos, M. (2017). *Preparing Microwave Transport Network for the 5G World*. Nokia.

31 Huawei (2016). *5G Network Architecture: A High-Level Perspective* (white paper). Huawei.

32 Huawei (2016). *5G Network Architecture* (white paper). Huawei.

33 Ericsson, "Intelligent Transport in 5G (Ericsson Technology Review)," Ericsson, 2017September.

34 5G Protocol Stack; 5G Layer 1, 5G Layer 2, 5G Layer 3. *RF Wireless World* http://www.rfwireless-world.com/Terminology/5G-Protocol-Stack-Layer-1-Layer-2-and-Layer-3.html. Accessed 1 July 2018.

35 Verizon, TS V5G.300," Verizon 5G Technical Forum, 01 07 2018. http://5gtf.net. [Accessed 1 July 2018].

36 Verizon, "2018 TS V5G.201," Verizon 5G Technical Forum. http://5gtf.net. Accessed 1 July 2018.

37 S. J. Syed, 2017 "Broadcasting over LTE/4G: Can it play a role in IOT & 5G?," NetManias. https://www.netmanias.com/en/post/blog/11731/5g-iot-lte-lte-b-embms/broadcasting-over-lte-4g-can-it-play-a-role-in-iot-5g. Accessed 22 June 2018.

7

Services and Applications

7.1 Overview

This chapter outlines expected services and application types for the 5G era. In addition to the enhanced functionalities serving smart devices and other equipment of consumers, 5G networks will be remarkably much more optimized for supporting extremely big amount of Internet of Things (IoT) devices. The machine-to-machine type of communications are expected to seriously take off well before the first International Telecommunication Union (ITU)-compliant 5G networks will appear in the markets, so there is a great need for tackling the challenging requirements of simultaneously connected devices – someday, there may be tens or hundreds of thousands of highly dynamic IoT devices in the field increasing the current signaling.

The following sections detail the expected 5G network procedures for supporting such a growing parallel communication. The network services need to cope with the requirements per each main category, i.e. building block of the 5G system. These can be generalized as follows:

- Evolved Multimedia Broadband communications (eMBB)
- Massive Internet of Things communications (mIoT)
- Critical communications (CriC)
- Network operations
- Vehicle-to-vehicle communications (V2V)

Some of the key references of the 3rd Generation Partnership Project (3GPP) utilized in this chapter are:

- *TS 23.501*. System architecture for the 5G system, stage 2
- *TS 23.502*. Procedures for the 5G system, stage 2
- *TS 23.503*. Policy and charging control framework for the 5G system, stage 2

The 3GPP Release 15 is the reference for the first phase 5G. The preparation for 5G era had already begun well before the freezing of Release 15 on June 2018, with late drops included through the end of 2018. The whole idea of 5G is to offer vastly expanded and enhanced services so the new generation will be considered as technology enablers for the respective new services and their markets. The 3GPP Release 14 prepared this ground by presenting multiple 5G research results.

5G Explained: Security and Deployment of Advanced Mobile Communications, First Edition.
Jyrki T. J. Penttinen.
© 2019 John Wiley & Sons Ltd. Published 2019 by John Wiley & Sons Ltd.

For the requirements of the 5G as seen by 3GPP, the key specification has been Technical Report TR 22.891, which details the feasibility studies on new services and markets technology enablers. This document consists of dozens of use cases that serve as a basis for a set of possible and expected opportunities in 5G telecommunications. Some examples of these use cases are related to the advanced IoT, vehicular communications (V2V and vehicle-to-everything [V2X]), and tactile Internet. The selected use cases represent a variety of requirements that 5G needs to fulfill in terms of capacity, coverage, latency, data speed, and reliability.

The previous mobile generations have served the end users by offering a uniform base for the communications. In the new era, 5G will optimize the network resources by offering different characteristics of the connections by introducing the concept of network slicing. In practice, network slices refer to virtualized subnetworks within the mobile network operator (MNO) radio coverage. Each slice can be adjusted to offer the best performance for the selected use cases. As an example, one slice can offer a vast simultaneous connectivity for low-bit-rate IoT devices whereas other slice can offer the lowest latency for real-time applications. These slices can be balanced by adjusting key performance and functional parameters, and thus the slices are the way to optimize also the respective costs of the connectivity. Not all the applications require expensive, high availability.

In addition to the network slicing, the above-mentioned 3GPP study also addresses many other aspects of the 5G networks to improve the performance. These include the support of variety of access technologies apart from the native 5G and their joint functionality, network flexibility aspects, and resource efficiency, among many other aspects. The outcome was the presentation of multiple service dimensions.

7.2 5G Services

7.2.1 General Network Procedures

As described in Chapter 4, 3GPP defines many completely new 5G elements and interfaces, in addition to the elements that cope with modified functionalities compared to the previous generations. Nevertheless, the 4G capacity – both from the radio's and core network's perspective – can also be allocated for the 5G users, which is especially beneficial in the initial phase of the deployments. This ensures a smooth adaptation to the new radio in such a way that the user can enjoy some of the benefits of the new generation. These deployment scenarios have been agreed in 3GPP, and they are summarized in the 5G architecture Chapter 4 of this book.

The 5G, in addition to a complete new set of network elements and virtualized, service-based architecture models, renews also the respective procedures for the different states of the network services. These procedures include the following:

- Registration management
- Service request
- User equipment (UE) configuration update
- Reachability
- Access network (AN) release

- N2
- Feature-specific UE and radio access network (RAN) information and compatibility

The states and procedures are summarized in the following sections, based mainly on the technical specifications for the stage 2 procedures for the 5G System (3GPP TS 23.502) and stage 2 system architecture for the 5G system (3GPP TS 23.501).

7.2.2 5G States

The 3GPP defines the signaling flows for the 5G connections, including initial attachment to the network referred to as registration, call establishment, mobility management (MM) procedures in idle and connected modes, packet data unit (PDU) session establishment, termination of the sessions, and deregistering from the network. Also, the detach procedure comes into the question when the user switches off the equipment, or it runs out of the battery.

Figure 7.1 summarizes the state diagram of 5G including a comparison with the 4G Enhanced Packet System (EPS).

In 5G system, the MM state of a UE can be either MM-REGISTERED or MM-DEREGISTERED. This depends on whether the UE is registered in 5G core

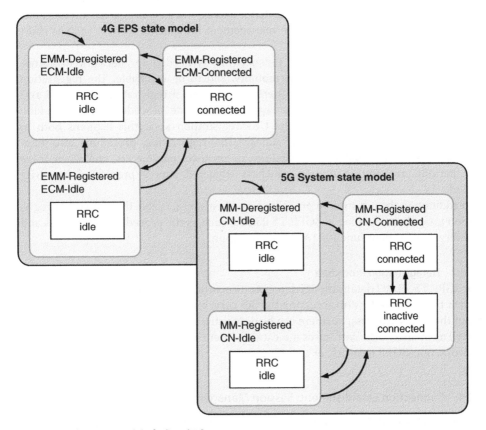

Figure 7.1 The state model of 4G and 5G.

network. As soon as the UE is registered, its state is CN-IDLE or CN-CONNECTED, depending on the presence of non-access stratum (NAS)-layer connection.

The abovementioned procedures and states are comparable with the 4G making the 5G MM and Core Network (CN) states similar to the EPS Mobility Management (EMM) and EPS Connection Management (ECM) states in the EPS. The MM-REGISTERED and MM-DEREGISTERED states of 5G are also similar to the 4G EMM principles. Nevertheless, the Radio Resource Control (RRC)-INACTIVE CONNECTED state of 5G is renewed for the RRC state model, to be applied as a primary sleeping state prior to RRC-IDLE state. As the UE enters the new state, the UE and RAN will keep the context information of the RRC connection for the UE. Examples of the data are UE capabilities and security context that are already existing as a result of the RRC connection setup, making the transition more fluent from inactive to active data transmission (3GPP TR 23.799, Study on Architecture for Next Generation System, V.1.2.1 [Nov. 2016]).

7.2.3 Registration and Mobility Management

7.2.3.1 Registration Procedure

The registration and deregistration procedures provide the functionality to register or deregister a UE with the 5G system, together with additional functionality to support also registration management for non-3GPP access. There is also a set of additional functionalities to support registration management for specific services such as short message service (SMS) over NAS [1].

The registration of the UE with the network is needed for network to authorize it to receive services, to enable mobility, and to be reachable. The registration procedure takes place when the UE performs the initial registration to the 5G system, or when it executes a mobility registration update procedure as the tracking area (TA) changes differing from the UE's registration area (this happens both in CM-CONNECTED and CM-IDLE states). The registration procedure also happens when the UE updates its capabilities or protocol parameters, and when there is a periodic registration update, which is triggered after a predefined inactivity period.

In addition to the registration procedure case just summarized, there are other registration procedures detailed in 3GPP TS 23.502, such as the procedure with access and mobility management function (AMF) reallocation.

7.2.3.2 Deregistration Procedure

The 3GPP TS 23.502 details the 5G deregistration procedure [1]. It is designed for UE to inform the network to stop accessing the 5G core, and the 5G core to inform the UE about the released access. So, there can be deregistration procedure initiated by the UE as well as the network. Both cases indicate whether the deregistration applies to 3GPP network, non-3GPP network, or to both cases.

7.2.4 Connection Establishment: Session Management

The 3GPP TS 23.501 details various procedures for the 5G system in Release 15, including session management [1]. The following summarizes some key procedures.

7.2.4.1 Session Management Procedures

The 5G session management procedures involve the reference point, i.e. interface *N4* for the respective interactions.

The 3GPP TS 23.501 considers cases for home-routed (HR) PDU Session and HR in roaming, and respective NAS SM (session management) information flow, which the session management function (SMF) processes.

For the HR PDU session, the information flow includes:

- For visited SMF (V-SMF) and home SMF (H-SMF), there is indication of NAS SM signaling for the PDU session establishment, PDU type, UE's IPv4 address and interface identifier, and session's Aggregate Maximum Bit Rate (AMBR).
- In addition, relayed by V-SMF but processed only by H-SMF, there is information on the Session and Service Continuity (SSC) mode, options for protocol configuration, quality of service (QoS) rules, and SM PDU DN request container.
- The H-SMF provides the V-SMF with the IPv6 prefix for the respective PDU session for facilitating data storage in visited country.

As detailed in 3GPP TS 29.502, for the HR roaming scenarios, the information includes:

- The NAS SM information that the V-SMF transfers in a container to H-SMF, in case V-SMF does not need to process it, or if it does not comprehend it due to unknown information element.
- The NAS SM information that the H-SMF transfers to V-SMF container.

7.2.4.2 PDU Session Establishment

The 3GPP TS 23.502 details UE-requested PDU session establishment for nonroaming and roaming with local breakout. In 5G, the UE-initiated PDU session can be initiated in the following cases:

- PDU session establishment procedure
- PDU session handover procedure between 3GPP and non-3GPP
- PDU session handover from EPS to 5GS

The network-initiated PDU session establishment procedure refers to a case where the network delivers a triggering message to the application of the device.

7.2.4.3 PDU Session Modification

The PDU session modification procedure takes place when any quality of service (QoS) parameter sent between the UE and network is modified. The respective details are presented in the 3GPP TS 23.501. The UE and network requested PDU session modification procedure includes means for both nonroaming and roaming cases with local breakout.

7.2.4.4 PDU Session Release

As the data transmission ends, the PDU session release procedure takes place to liberate the respective resources. These resources include the Internet Protocol (IP) address and prefixes allocated for the PDU session, and user plane function (UPF) resources of this PDU session such as *N3* and *N9* interface termination.

The SMF informs all the entities involved in that specific PDU session, including the data network and policy control function to indicate that the PDU session is released.

7.2.4.5 Session and Service Continuity

The change of the PDU sessions triggers a procedure that is managed by SMF for changing the PDU session anchor. The principle of this procedure is to release the previous PDU session associated with an old PDU session anchor and to immediately establish a new PDU session with a new PDU session anchor to the same data network.

7.2.5 IMS Calls

5G supports IP Multimedia Subsystem (IMS) calls. They are defined in 3GPP TS 23.501, Ch 4.4.3, and TS 23.502, Ch 4.13.6.

7.2.6 Short Message Service

The specific scenarios of SMS over NAS are described in 3GPP TS 23.501, Ch 4.4.2, and procedures for SMS over NAS are found in TS 23.502, Ch 4.13.3.

The mobile-originated (MO) and mobile-terminated (MT) SMS over NAS, by relying on either 3GPP or non-3GPP network, has different scenarios. If the security context is already activated prior to the transfer of the short message, the NAS transport message needs to be ciphered and integrity protected. It is based on the UE and AMF NAS security context as described in European Telecommunications Standards Institute (ETSI) 133 501, V.15.1.0, Chapter 6.4.

7.2.7 Public Warning System

The Public Warning System (PWS) is detailed in 3GPP TS 23.501, Ch 4.4.1, and TS 23.502, Ch 4.13.4.

7.2.8 Location-Based Services

The location-based service (LBS) is detailed in 3GPP TS 23.502, Ch 4.13.5.

7.2.8.1 Positioning Techniques in 5G

The 3GPP TS 38.305 details the stage 2 functional specifications of UE positioning in Next Generation Radio Access Network (NG-RAN) as of the Release 15. This specification allows the NG-RAN to utilize one or more positioning methods at a time to locate the UE. The positioning is based on signal measurements and position estimate. There is also an optional functionality to estimate velocity of the UE.

The UE, serving Next Generation evolved NodeB (G) (ng-eNB) or 5G NodeB (gNB), performs the measurements that might include E-UTRAN LTE and other, general radio navigation signals, e.g. via Global Navigation Satellites System (GNSS). The estimation of the position may be performed by the UE or by the Location Management Function (LMF). The positioning may rely on UE-based, UE-assisted, and LMF-based, or NG-RAN node assisted methods.

The 5G system's NG-RAN includes the following standard positioning techniques for the estimation of the UE's location:

- Network-assisted GNSS
- Observed time difference of arrival (OTDOA)
- Enhanced cell ID (E CID)
- Barometric pressure sensor positioning
- Wireless Local Area Network (WLAN) positioning
- Bluetooth positioning
- Terrestrial Beacon System (TBS)
- Hybrid positioning combining some of the above-mentioned methods
- Standalone mode that refers to an autonomous functionality of the UE without assistance from network, based on one or more of the above-mentioned methods

Table 7.1 summarizes the principle of the 5G positioning techniques.

7.2.8.2 5G Positioning Architecture

Figure 7.2 depicts the 5G positioning architecture for NG-RAN and E-UTRAN as described in the 3GPP TS 38.305 [1, 2]. As can be interpreted from the figure, the main element for location-based services is the LMF. The RAN may include an ng-eNB, gNB, or both. If both types of nodeBs are present, the communication further to the AMF takes place only by one respective interface.

In 5G context, a transmission point (TP) refers to a TP that is a set of geographically co-located transmit antennas for a single cell, a part of a single cell, or a single PRS-only. The PRS refers to a Positioning Reference Signal for E-UTRA. Transmission points can be formed, e.g. by antennas of ng-eNB or gNB, remote radio heads, remote antenna of a base station, or antenna of a PRS-only TP. PRS-only TP refers to a TP that solely transmits PRS signals for PRS-based TBS positioning for E-UTRA and that is not associated with a cell.

A single cell can consist of one or multiple transmission points. In practical deployment, each transmission point may correspond to one cell. Also, a single ng-eNB may be able to control several TPs, such as radio heads.

A UE-specific location-based request may be triggered by different entities such as the target UE or GMLC (Gateway Mobile Location Center). This request is routed to the AMF. Please note that the AMF itself can also trigger the location-based service request for a target UE. This can happen, e.g. when the UE triggers an IMS emergency call.

Following the architecture model of Figure 7.2, the AMF sends a location services request to LMF for its processing, possibly accompanied by related assistance data. After the LMF has processed the data and resolved the UE's location, it sends the results back to the AMF.

The LMF may have further signaling connection to E-SMLC (Evolved Serving Mobile Location Center), which is defined in 3GPP TS 29.516, and/or SUPL (Secure User Plane Location) platform. The E-SMLC provides means for an LMF to access information from E-UTRAN to, e.g. support the OTDOA for E-UTRA positioning. The SLP, in turn, is a SUPL entity that is applicable for positioning over the user plane.

Table 7.2 summarizes the tasks of the functional elements for the 5G positioning procedures. Table 7.3 summarizes the interfaces of Figure 7.2.

Table 7.1 The positioning techniques of the 5G system for locating UE.

Method	Principle
Network-assisted GNSS	UE can receive GNSS signals, e.g. from global positioning system (GPS) variants, Galileo, GLONASS, SBASs (Space Based Augmentation Systems), QZSS (Quasi Zenith Satellite System) and BDS (BeiDou Navigation Satellite System). Different GNSSs can be used separately or in combination.
OTDOA	UEs measure timing of received downlink signals from various TPs, comprising eNBs, ng-eNBs, and PRS-only TPs. The UE gets assistance-data from the positioning server. The location of the UE is relative to the neighboring TPs.
Enhanced Cell ID	The basic Cell ID (CID) estimates the UE's position based on serving ng-eNB, gNB and cell information, which can be obtained, e.g. by paging and registration. Enhanced Cell ID (E CID) improves the accuracy based on additional UE measurements, NG-RAN radio resource and other measurements. For serving gNB, E CID can use E-UTRA measurements originated from UE to the serving gNB.
Barometric pressure sensor	Barometric pressure sensors of the UE determine the UE's vertical position. This data may be optionally aided by other data for measurements and to send measurements to the positioning server for further estimate, and the method is combined with other positioning methods to determine the 3D position of the UE.
WLAN	UE's location can be obtained based on WLAN's access point (AP) identifiers and databases as the UE measures signals from WLAN Aps, combined with optional assistance data, and sends the measurements to the positioning server, which calculates the UE's location based on these results and a references database. The UE may also determine its location using WLAN measurements and AP assistance data originated from the positioning server.
Bluetooth	The UE's location is determined via its measurements of Bluetooth beacon identifiers, supported optionally with other measurements. The UE's location can be obtained based on these measurements and a references database, the location of the UE, is calculated.
TBS	The UE's location is determined as it measures TBS (Terrestrial Beacon System) signals, combined with optional assistance data. The UE may send measurements to the positioning server for further analysis. The TBS is a network of terrestrial transmitters dedicated for positioning services, and it has two variants: MBS (Metropolitan Beacon System) and Positioning Reference Signals (PRS). The latter is part of OTDOA positioning.

7.2.9 Dual Connectivity

The 3GPP TS 23.502 details the support of dual connectivity (DC) in 5G. This presented example is related to RAN-initiated QoS flow mobility, which refers to the case for transferring QoS flows between master and secondary RAN node based on IP connectivity. In this case, the SMF and UPF remains the same. If QoS flows are present for multiple PDU sessions, the same procedure is repeated for each individual PDU session (Figure 7.3).

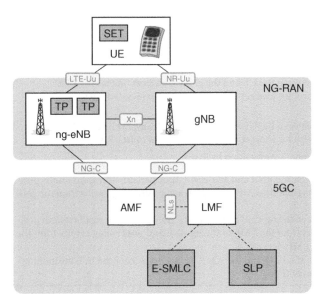

Figure 7.2 5G positioning architecture. There is only one next-generation core (5G) (NG-C) present when both ng-eNB and gNB are present.

Table 7.2 Functions of the 5G elements for UE positioning.

Functional element	Description
UE	The user equipment may perform downlink radio measurements of NG-RAN, E-UTRAN, GNSS, TBS, WLAN access points, Bluetooth beacons, and UE barometric sensors, and LCS applications residing in the UE or network. The UE may also contain independent location services such as GPS.
gNB	5G nodeB provides measurement information for a target UE and to an LMF.
Ng-eNB	Enhanced 4G nodeB measures and provides measurement results for a target UE and to an LMF. Performs measurements upon LMF's request. May serve various TPs.
LMF	The location management function supports location services for target UEs. Manages positioning of UE and delivers assistance data to UE. Interacts with serving gNB and/or ng-eNB to get position measurements for target UE, including ng-eNB uplink measurements and UE downlink measurements, which are done during other procedures such as handover. The LMF delivers assistance data upon request. The LMF decides on the applied position method, or set of them in hybrid mode, based on criteria such as LCS Client type, QoS, UE's, gNB's, and ng-eNB's positioning capabilities.

7.3 Network Function-Related Cases

7.3.1 Network Exposure

The 5G network exposure functionality is defined in 3GPP TS 23.502, Chapter 4.15.

Table 7.3 The interfaced in 5G system for positioning services.

Interface	Description
NR-Uu	Radio interface between the UE and the gNB. Transport link for the LTE Positioning Protocol for a target UE with New Radio (5G) (NR) access to NG-RAN.
LTE-Uu	Radio interface between the UE and the ng-eNB. Transport link for the LTE Positioning Protocol for a target UE with LTE access to NG-RAN.
Xn	This interface is under investigation by 3GPP.
NG-C	1) Interface between the gNB and the AMF: transparently transports both positioning requests from the LMF to the gNB and positioning results from the gNB to the LMF. 2) Interface between the ng-eNB and the AMF: transparently transports both positioning requests from the LMF to the ng-eNB and positioning results from the ng-eNB to the LMF.
NLs	Interface between the LMF and the AMF. Transport link for the LTE Positioning Protocols LPP and NRPPa.

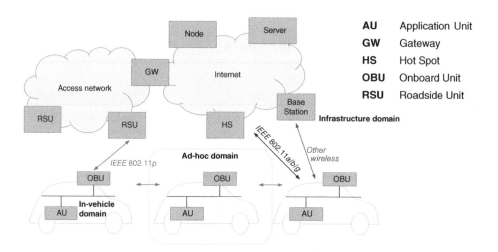

Figure 7.3 "Traditional" architecture of V2X based on IEEE 802.11p and other variants.

7.3.2 Policy

The 5G policy scenarios are defined in 3GPP TS 23.502, Chapter 4.16.

7.3.3 Network Function Service Framework

The network function (NF) service framework is defined in 3GPP TS 23.501, Chapter 4.17.

7.4 Vehicle Communications

One of the pillars for 5G is vehicle communications. In fact, 5G provides a logical platform for large areas, and thus, the V2X served by 5G offers optimized methods to tackle the communications between the vehicles and rest of the infrastructure.

7.4.1 How V2V Works

The cooperative intelligent transportations system (ITS) and car-2-car communication is based on the IEEE 802.11p WLAN at 5.9 GHz in Europe and in the United States. When two or more vehicles (ITS stations) appear within the radio communication range, they connect automatically and establish an ad-hoc network where all ITS stations know the position, speed, and direction of the other stations and will be able to provide warnings and information to each other.

A single WLAN link is limited to about 1000 ft, or a few hundred meters. Each vehicle also works as a router and may deliver messages via multiple hops to other vehicles and ITS stations. The routing is adaptive and can cope with fast changes within the ad-hoc network topology.

Drivers benefit from Cooperative Intelligent Transport System (C-ITS) support, e.g. when approaching the end of a traffic jam, if a vehicle in front suddenly brakes hard, and if a broken-down vehicle or road construction blocks the route [3]. Applications of V2V include forward collision warning, blind spot warning and lane change warning, emergency brake light warning, control loss warning, and no pass warning [4]. There is active research in vehicle to infrastructure (V2I) on signal phase and timing (SPaT). It is related to the ability to coordinate driving speeds with traffic light patterns, and to maximize fuel economy and velocity. SPaT patterns may optimize traffic flow in high-traffic areas and synchronize the traffic according to the traffic light patterns.

7.4.2 V2V System Overview

A V2V communication system comprises three distinct domains: in-vehicle; ad-hoc; and infrastructure domain.

7.4.2.1 In-Vehicle
The in-vehicle domain refers to a network logically composed of an on-board unit (OBU) and application units (AU). The AU refers to a dedicated device that executes a single or a set of applications and utilizes the OBU communication capabilities. The AU can be an integrated part of a vehicle and can be permanently connected to the OBU. It can also be a portable device such as a laptop that can dynamically attach to the OBU. The AU and OBU are typically connected via wire, but the connection can also be wireless, such as Bluetooth.

7.4.2.2 Ad-hoc
The ad-hoc domain, or vehicular ad-hoc network (VANET), is composed of vehicles equipped with OBUs and stationary units along the road, i.e. roadside units (RSUs). The OBU is equipped with a short-range wireless communication device dedicated for road safety, and potentially with other optional communication devices.

The OBUs form a mobile ad-hoc network (MANET) for communications among nodes in a distributed manner without the need for centralized coordination instance.

7.4.2.3 Infrastructure

While RSUs for Internet access are typically set up with a controlled process by a Car-to-Car Communication Consortium (C2C CC) key stakeholder (such as road administrator), public or privately owned hot spots (HSs) are usually set up in a less-controlled environment. These two types of infrastructure domain access, RSU and HS, also correspond to different applications types.

If neither RSUs nor HSs provide Internet access, OBUs may utilize communication capabilities of cellular radio networks (2G, 3G, 4G, 5G) if they are integrated in the OBU.

The infrastructure domain is linked to a public key infrastructure (PKI) certification infrastructure. The certification authority (CA) issues digital certificates to OBUs and RSUs.

Please note that some sources interpret vehicle-to-roadside (V2R) as its own communication domain (representing a special case of V2I).

7.4.3 V2V Reference Architecture

The "traditional" architecture of V2X is based on IEEE 802.11p connectivity. ITS America supports the US Department of Transportation's (USDOT) connected vehicle program that is focused on the use of dedicated short-range communications (DSRC)/wireless access in vehicular environments (WAVE). USDOT has constructed a pilot to demonstrate effectiveness of DSRC with a mix of light, heavy, and transit vehicles.

Nevertheless, cellular-based, and especially 3GPP LTE-based connectivity is becoming more feasible for the local vehicle communications, too, along with the further standardization of, e.g. direct communications between the user devices.

5GAA (5G Automotive Association) is a rather new initiative to drive for the cellular connectivity in V2X environment. The forum represents the automotive industry, and there are also many other companies as members, including Verizon and Gemalto.

7.4.4 Development of V2X

V2X evolves and will include enhanced concepts in Advanced Driver Assistance Systems (ADASs). ADAS refers to a system that provides means for vehicles to cooperate, coordinate, and share sensed information. V2X will grow into Connected and Automated Driving (CAD).

As soon as 5G networks and corresponding reliable, low-latency, and mission-critical services are available for V2X applications, ADAS and CAD can be enhanced further.

V2V communication market size was at more than USD15 billion in 2015 and is likely to grow at an estimated annual rate of more than 5% by 2023 [4]. Connected vehicles may generate significant data, which brings new market potential for the data owner. It also can be assumed that the developing ecosystem provides new business opportunities to the involved and to third parties.

7.4.5   Security

V2V security design benefits from advanced OTA capabilities in cooperation with intelligent methods protecting against threats in order to ensure both occupant safety and vehicular serviceability.

In North America, the USDOT National Highway Traffic Safety Administration (NHTSA) has issued a Notice of Proposed Rulemaking (NPRM) to mandate vehicle-to-vehicle communication technology for new light vehicles in the United States.

7.4.6   NHTSA Requirements

The NHTSA is proposing to issue a new Federal Motor Vehicle Safety Standard (FMVSS) No. 150, to require all new light vehicles to be capable of V2V communications, such that they will send and receive basic safety messages (BSMs) to and from other vehicles.

The document detailing the rules is "NHTSA 49 CFR Part 571, Docket No. NHTSA-2016-0126, RIN 2127-AL55, FMVSS; V2V Communications."

The proposal contains V2V communication performance requirements predicated on the use of on-board DSRC devices to transmit BSMs about a vehicle's speed, heading, brake status, and other vehicle information to surrounding vehicles, and receive the same information from them. This would help vehicle systems identify potential crash situations with other vehicles and warn their drivers.

7.4.7   IEEE 802.11p vs. LTE-Based Short-Range Communications

7.4.7.1   Background

The V2X communications has been "traditionally" based on the IEEE 802.11p connectivity in such a way that the vehicles as well as the roadside elements automatically establish the data transmission in an ad-hoc network. As the 3GPP LTE standards are evolving, they also start including functionalities that can be utilized in vehicle communications, such as direct connection between the user equipment.

5GAA is driving for the cellular variant to be included in the V2X. 5GAA is emphasizing the enhanced performance of the LTE-based dedicate short-range communications over not-so-optimal IEEE variant. As an example, C-V2X (cellular vehicle-to-everything) is straightforward via the LTE itself while the WLAN requires hot spots in surrounding areas to connect beyond the direct data transmission between vehicles.

Benefits of C-V2X over IEEE 802.11p include:

- Synchronous vs. asynchronous enhances spectral efficiency as it enables time division multiplexing and lowers channel access overhead.
- Frequency Division Multiplexing (FDM) and Time Division Multiplexing (TDM) vs. solely TDM optimizes radio link budget and thus cell range.
- Turbo coding vs. convolutional coding provides a gain of approximately 3 dB, which benefits the radio link budget.
- HARQ (Hybrid Automatic Repeat and Request) vs. no HARQ further enhances the cell range.

- Single-Carrier Frequency Division Multiplex (SC-FDM) vs. Orthogonal Frequency Division Multiplexing (OFDM) optimizes the power amplifier's transmission power.
- Semi-persistent transmission and relative energy-based selection vs. CSMA-CA (Carrier Sense Multiple Access and Collision Avoidance) optimizes the resource selection.

According to 5GAA, while C-V2X direct services can technically coexist in adjacent channels with IEEE 802.11p-based radio access, the cellular can address all V2X applications in an end-to-end manner with the same technology and is therefore scalable and evolvable. As part of the 3GPP standards, C-V2X offers an evolution path from LTE to 5G. This is why 5GAA reasons that C-V2X (3GPP-defined cellular for V2X environment) is a more logical choice than hybrid solution that incorporates IEEE 802.11p-based technology. 5GAA recognizes that C-V2X is future proofed, and has superior direct communications performance compared to 802.11p. It also offers a higher degree of security for all operating modes via embedded security for V2N transmission and the equivalent PKI security services defined in standards using 802.11p. 5GAA thus favors C-V2X for V2V, V2I, V2P, and V2N.

7.4.7.2 Comparison

According to [5], there are two main use cases for vehicle communications: telematics and V2V/V2I ITS road safety.

Telematics is characterized by infrequent transmission, large coverage, and latency insensitivity, while underlying technologies are cellular networks (2G/3G/4G). *V2V/V2I ITS road safety*, in turn, is characterized by frequent transmission (1–10 Hz), short-range communication ($n \times 100$ m), low latency requirement (<100 ms), and high reliability, while underlying technologies are 802.11p and LTE-V2X.

Ref. [5] notes that IEEE 802.11p performance cannot be guaranteed due to its CSMA/CA based ad-hoc mechanism, with hidden terminal and congestion problems. Its spectrum dedicated for V2V road safety is limited (10 MHz for United States, 30 MHz for Europe), and it is rather expensive to deploy RSUs to cover telematics services. Furthermore, its business model is not clear, and there are security and maintenance issues (downloading, update, revocation, etc.). There is thus no clear evolution path for future services via the IEEE 802.11p path.

7.4.8 Security Threats

As some examples, potential attacker can intend to jailbreak a V2V communication system to send arbitrary messages to other vehicle, including false alarms to drivers. There could thus be messages that exploit the car system to access the physical capabilities of the V2V system. The BSM is the primary message set proposed to send data between vehicles and between vehicles and the infrastructure.

Although the BSM is mainly developed for safety applications, it may also be used by other connected vehicle applications, such as mobility, weather, and AERIS (Applications for the Environment: Real-Time Information Synthesis) programs, delivering safety benefits, too. Currently, USDOT is developing the applications. The assurance of the secure communication of these messages is essential. Also, as for congestion, it is critical that safety messaging not be compromised due to broadcasting more data for V2I.

7.4.9 MNO Role in V2V

3GPP is actively standardizing V2V functionality in LTE. Some examples of the contributions are, based on the Ref. [6]:

- S1-151034 V2V value for high-speed requirement, Huawei
- S1-151035 V2V clarification that comm. range is distance between UEs, Huawei
- S1-151036 V2V loss requirement clarification, Huawei
- S1-151037 V2V service requirement values, Huawei
- S1-151103 V2I V2V communication, LG
- S1-151061 V2V terminology update on V2X enabled, vehicular UE, etc., Huawei
- S1-151071 V2P enhanced tethering, Sony
- S1-151091 V2I traffic flow optimisation use ccase, Nokia Networks
- S1-151093 Proposed plan for V2X (Stage 1), LG Electronics
- S1-151171 V2V service requirements out of coverage, Huawei
- S1-151179 V2X use case: Traffic congestion control/warning, KT Corp.

The 3GPP SA1 group has agreed 18 use cases in the technical recommendation TR 22.885. Some examples of these are:

- V2V safety
- V2V/V2I traffic efficiency
- V2V message transfer under operator control
- V2I Road safety
- V2P Road safety
- V2X in areas outside network coverage

One of the main roles of MNO in V2V business is the ability to control V2V communication, which refers to the ability of MNO to perform authorization, session establishment, and resource allocation. Furthermore, the RSU entity can be implemented either in an evolved NodeB (eNodeB) or a stationary UE. Standardization work has been active across 3GPP. The supporting LTE-based technology is suitable for V2X services and will further evolve for future V2X. The new cellular network assisted LTE-D2D provides a logical business opportunity for all C-ITS stakeholders.

The 5GAA has taken the role of driving the technology adaptation in a form of an industry forum. It is beneficial to promote the cellular-based technology among car makers, MNOs, road operators, authorities, for standard, spectrum, deployment, and business model definition.

To date, USDOT has been reluctant to consider cellular-based telematics because of concerns that cellular carriers will not commit to a single technology (2G, 3G, or 4G consistently nationwide etc.) over a long period of time, forcing car owners and state DOTs to constantly upgrade equipment in vehicles and intersections, respectively.

Telematics service providers, infrastructure ITS services (e.g. tolling), M2M, and other application service providers are thus waiting for more information to commit to correct technology. Road operators and vehicle manufacturers need to examine future WAN services critically, based on assessment of their potential longevity and their cost, service quality, and coverage.

USDOT is interested in 4G LTE-Advanced and other longer-range systems as an option to support the connected vehicle system. USDOT will publish reports on

LTE-Advanced and on telematics/M2M, both authored by ITS America. There is thus an opportunity for LTE and 5G to support a connected V2V program, as well as other vehicle telematics and M2M applications.

7.4.10 Markets for Peer-to-Peer and Cellular Connectivity

The two modes, peer-to-peer (IEEE variant) and cellular V2V (3GPP variant), are at least partially competing technologies. The benefit of the IEEE 802.11p solution is the established technology, but it is limited in its Wi-Fi coverage area, including the connectivity between vehicles and roadside or pedestrian units. The benefit of a 3GPP solution is the wide coverage while it is still emerging technology that has not established market position.

US DoT anticipates that a foundational network (so-called network of networks) will need to be developed to exchange diagnostics, security, and authentication data between vehicles and a V2V centralized cloud-based network management authority to manage V2V applications nationwide. Options include a network of DSRC or Wi-Fi hot spots at intersections across the country, plus possible utilization of WANs such as cellular and satellite. As 5GAA is actively promoting the cellular-based DSRC system, it may become relevant option.

In both cases, there is room for connectivity and value-added services offered by MNOs and other stakeholders. As many IoT devices are located in remote areas, such as power meters, the provisioning of the subscriptions of such equipment is needed. This might also apply to the automotive industry, depending on the use cases.

V2V technologies will become more important in the near future. The designed architecture and functionality will be enhanced even as the system will be implemented in new environments. Likewise, the importance of the respective security increase as for the communications links internally within the vehicle, between different vehicles as well as with external entities. The technology is going to be increasingly popular along with the new critical communications technologies such as self-driving cars. As 3GPP has been active in developing the foundation for IoT via the cellular networks, also the 5G could play a major role in the vehicle communications.

7.5 Machine Learning and Artificial Intelligence

As 5G takes off, concrete steps can be taken in the development of artificial intelligence (AI), which can assist in many areas on top of 5G services. It can be generalized that the evolution along with 5G deployments are moving the "traditional" models, relying on autonomic network management to cognitive network management. The transition includes machine-learning abilities combined with self-aware, self-configuring, self-optimizing, self-healing, and self-protecting networks.

References

1 3GPP, "TS 23.502 V15.1.0, Procedures for the 5G System; Stage 2 (Release 15)," 3GPP, 2018.

2 3GPP, "TS 33.305 NG-RAN Positioning," 3GPP, 2018April.

3 CAR 2 CAR Communication Consortium, "Overview of the C2C-CC System," C2C CC, 2007August.

4 Global Market Insights, "Vehicle to vehicle (V2V) communication market size," Global Market Insights, 2018. https://www.gminsights.com/industry-analysis/vehicle-to-vehicle-v2v-communication-market. [Accessed 26 July 2018].

5 Shi, Y. (2015). *LTE-V: A Cellular-Assisted V2X Communication Technology*. Beijing: Huawei.

6 3GPP, "3GPP TDocs (written contributions) at meeting S1-70, 2015-04-13 to 2015-04-17, San Jose del Cabo," 3GPP, 13 April 2015. http://www.3gpp.org/DynaReport/TDocExMtg--S1-70--31352.htm. [Accessed 26 July 2018].

8

Security

8.1 Overview

Cybersecurity – referring to threats and respective protection – is one of the most important topics in modern mobile communications. The increasing popularity of the smart devices has exploded the amount of malware that may lurk within apps, web pages and elsewhere in the code of both software and hardware of the devices, networks and systems. The current growth of the Internet of Things (IoT) business is also one of the potential targets of the vulnerability attacks.

This chapter presents special security aspects of 5G, giving first an overview on the current security up to 5G. This includes both internal and external threats.

The following sections go through the network security functions and elements. The 5G procedures and security processes are upgraded from previous systems, and there are plenty of new aspects within the network and in interoperable modes. This chapter thus also presents new key methods for security monitoring and enhanced Lawful Interception (LI).

Then, threats and protection in mobile communications, and specifically in 5G era, are discussed. This part deals with the security compromising trends that indicate the security breach types in 5G systems (5GS) and presents strategies for protection.

One of the foundations for the security of the mobile communications is related to the secure element (SE), so this chapter discusses the role of removable security elements and embedded SE in providing adequate 5G security. This part includes information on the removable SIM (subscriber identity module) and UICC (universal integrated circuit card), as well as other removable elements and embedded Secure Elements such as eUICC (embedded universal integrated circuit card). Devices without hardware-based secure element are discussed as of the relevant technologies for use cases relying on cloud and other nonsecure element solutions. These use cases include payment, access, and other solutions relevant for near-field wireless communications and extremely economic IoT devices lacking security.

Finally, enhanced security solutions are discussed, such as integrated universal integrated circuit card (iUICC), tokenization, trusted execution environment, digital rights management (DRM), host card emulation, and cloud.

Among other security-related specifications, the key documents related to this chapter are the 3rd Generation Partnership Project (3GPP) TS 33.899 (study on the security aspects) and TS 33.501 (5G security architecture).

5G Explained: Security and Deployment of Advanced Mobile Communications, First Edition.
Jyrki T. J. Penttinen.
© 2019 John Wiley & Sons Ltd. Published 2019 by John Wiley & Sons Ltd.

As the second phase of 5G will gradually be reality and the respective globally interoperable deployments will take off as of 2020, the security aspects are going to be more relevant than ever in wireless environment. With the steadily increasing news on the massive security breaches, it seems that sufficiently well protected environment is rather an exception than something to be taken of granted.

5G represents completely renewed architecture, technical functionality and radio interface, which means that there may be also new, unknown security holes despite of all the preparation. The evaluation of such security threats is of utmost importance in both consumer as well as IoT environments in order to design appropriate shielding. The protection mechanisms of the 5G system require a tailored approach for each usage environment, depending on the respective business model and the point of view of the involved stakeholders.

The completely new architectures and technologies of the 5G infrastructure, such as *virtualization of network functions (NFs), network slicing,* and *mobile-edge computing (MEC)* have major impact on the design of the security. As the value of the assets dictates the level of the technical requirements, the 5G security must be scalable, flexible, and adaptive per each use case category, both within the network as well as on the device.

This chapter discusses the 5G security, including new technical opportunities for 5G stakeholders in the assurance of the protection. [1] One of the security areas is related to the concept of evolving SIM, or universal subscriber identity module (USIM), which may address many security concerns in the communication end points, including wireless consumer and IoT devices. It is of utmost importance to protect valuable assets of the stakeholders, such as network subscription credentials within the user equipment (UE) by the means of dedicated and evolved SIM security solutions.

The evolved SIM concept, which is taking shape in the 5G era, must offer a variety of technical capabilities. It will therefore have to be scalable and provide optimal grade of security per each use case considering the respective assets to be protected. In other words, the optimal protection makes the cost of the malicious intentions higher than the value of the benefit of such act is.

In addition to the evolution of the traditional SIM/USIM, there are emerging technologies such as eUICC, iUICC, and vUICC (virtual universal integrated circuit card). Their adaptation into practice depends on the willingness of mobile network operators (MNOs) to move into that direction, assuming these concepts provide sufficient security. The development of both SW and HW of these technologies are of interest to many stakeholders, yet are mainly driven by chipset vendors. The concept of such solutions would change not only the technology itself but also the "traditional" way of work between the current stakeholders, i.e. chipset manufacturers, device manufacturers, SIM/USIM providers, and MNOs. The functional and certified trust model is of utmost importance in this new environment to guarantee that the new solutions are at least equally secure, if not more, than the previous UICC variants.

Up to now, the security of mobile networks has been based on the chain of trust in such a way that the MNOs may rely on the SIM card vendors as they are certified and comply with the strictest requirements for maintaining the secrets physically and logically via specific security processes. Figure 8.1 illustrates the traditional and new models of the chain of trust.

Along with the deployment of 5G systems, the role and ownership model of the security platform within the device makes it possible for the evolved SIM to serve as a

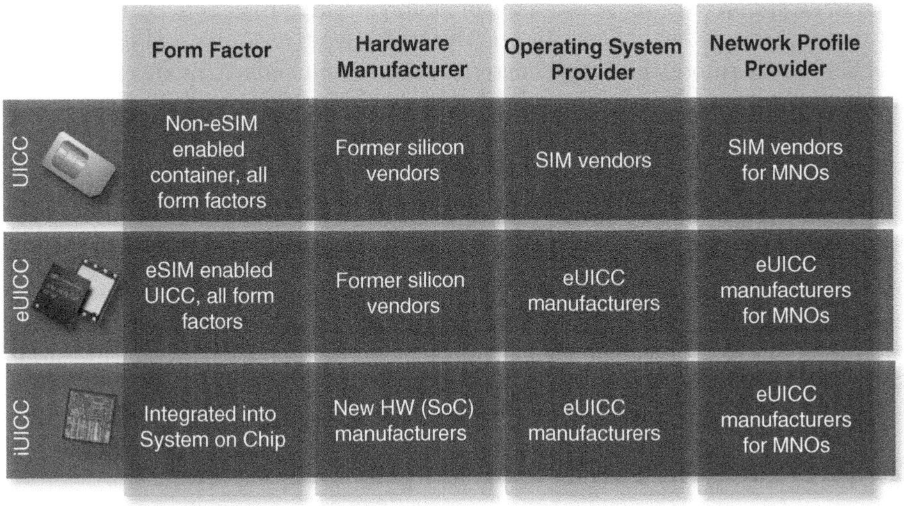

	Form Factor	Hardware Manufacturer	Operating System Provider	Network Profile Provider
UICC	Non-eSIM enabled container, all form factors	Former silicon vendors	SIM vendors	SIM vendors for MNOs
eUICC	eSIM enabled UICC, all form factors	Former silicon vendors	eUICC manufacturers	eUICC manufacturers for MNOs
iUICC	Integrated into System on Chip	New HW (SoC) manufacturers	eUICC manufacturers	eUICC manufacturers for MNOs

Figure 8.1 The chain of trust will evolve along with the new variants of the UICC.

security anchor for other stakeholders such as device manufacturers and application providers. These new ownership models could potentially make evolved SIM accessible for all the stakeholders requiring security in the end points.

Based on the future variants of the evolved SIM concept, which may consist of variety of security solutions such as embedded and integrated SIM for the end points, there might be a need for a flexible and virtualized security management server system. Some of the key technologies capable of handling the new demands are based on *device management* and over-the-air (OTA) platforms for the support of various *authentication mechanisms* and types of *security credentials* on network, application, device, and other layers. Those systems could be adapted to address the emerging market of closed networks that may be owned and operated by factories or enterprises.

For all the new environments formed by the closed networks, the management of the *identities* of vast amount of consumer and IoT devices is one of the most important development areas in providing the security. In addition to the consumer networks as well as typical IoT networks such as sensor networks, there are many other environments, too. One of these is cyber physical security (CPS), which refers to industrial machinery and the safe operation, maintenance, and telematic communications between authorized personnel and equipment. These solutions may be deployed to protect the network infrastructure, communication, and authentication between network entities. It is also possible to collaborate with technology providers such as network infrastructure providers and chipset makers to integrate evolved SIM and network components into adequate security solutions. To accelerate the offering, collaboration with MNOs is also beneficial.

In addition to the abovementioned topics, this chapter gives an overview to current security aspects relevant to 5G environment, including internal and external threats, and details network security functions and elements. The latter includes updated procedures and security processes from previous systems and aspects within the network and in interoperable modes, new methods for security monitoring, as well as enhanced

Lawful Interception (LI). There is discussion of threats and protection, i.e. the security compromising trends that indicate the security breach types in 5G era, and strategies for protection. Furthermore, this chapter describes the current understanding of the secure element, and more specifically the role of removable security element and embedded SE in 5G security, including SIM/UICC cards and removable memory cards and universal serial bus (USB). Another relevant 5G topic is the vast amount of IoT devices that partially will be deployed without secure element, so there is description of relevant technologies for use cases relying on cloud and other nonsecure element solutions, including payment, access, and other solutions relevant for near field wireless communications and extremely economic IoT devices lacking security. To finalize the 5G security considerations, this chapter discusses enhanced security solutions including novelty solutions based on embedded secure elements, tokenization, trusted execution environment, DRM, host card emulation, and cloud.

8.2 5G Security Threats and Challenges

8.2.1 Trends

Overall Internet development is leading the way to virtualization of practically all the communications of key businesses such as banking and governance. This has been a driving force for economic growth, as it clearly enhances the efficiency and ensures the optimal performance of the functions. There is no way back to the old days when, by default, people needed to queue everywhere to handle simple things like paying bills, making bank deposits, or buying groceries.

Along with easing that aspect of life, the Internet has demonstrated the vulnerability of the societies when relying completely on the virtualization. There are countless examples of how financial institutes, power transmission companies, and other entities representing critical infrastructure have been paralyzed for sufficiently large time laps for citizens to realize how fragile the current protection can be.

5G represents a major step in the mobile communications development, yet its stakeholders like MNOs, standardization bodies, and equipment manufacturers, have equally important role in ensuring the sufficient protection mechanisms. 5G is not designed only for consumer markets but is planned to support remarkable volumes of IoT traffic via mIoT use cases, i.e. massive Internet of Things. This may open further totally new doors for malicious intentions.

There is thus a special emphasis on the security aspects of 5G. The respective security standards detail many new functions in addition to the ones that are applied in the previous generations, enhancing further the performance under attacks such as DDoS (distributed denial of service) or MITM (man in the middle).

One of the documents summarizing the latest trends on the Internet and mobile communications environments is the annual Verizon Data Breach Investigations Report. It is a good reminder and base for designing and fortifying the respective security shielding.

The report divides the data incident patterns into categories: insider and privilege misuse (access via trusted actors), cyber-espionage (targeted external attacks), web application attacks (stolen credentials and vulnerability exploits), crimeware (malware incidents), point-of-sale intrusions (POS attacks aiming to payment card data disclosure), denial of service attacks (DoS or distributed DoS aiming to affect business

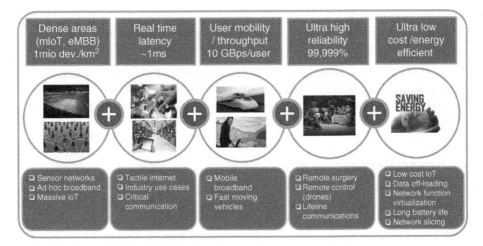

Figure 8.2 5G targets and use cases.

operations), payment card skimmers (tampering of ATMs and terminals), physical theft (loss of IT data), and miscellaneous errors (causing data loss).

As for the mobile communications point of view, the main categories of this report are related to crimeware, DoS and web attacks, although there are various other categories that can indirectly impact on the security via mobile communications systems, such as exploiting the weaknesses of payment via near-field communications (NFCs). One of the remarkable trends has been the heavy increasing of the ransomware, which is a valid threat in both the fixed and mobile environment.

8.2.2 Needs in Mobile Environment

5G aims to achieve significant improvements over existing mobile technologies along with the aim to enable new use and business cases. Figure 8.2 depicts the high-level key targets of 5G and lists typical use cases requiring the enhanced technical capabilities provided by the 5G networks. At present, the ideas for highly advanced use cases benefiting from 5G are merely forming, so it is highly possible that there will be totally new solutions seen as the 5G takes off. Further information on use cases and requirements can be found in [2].

While several high-level security requirements such as the need for identification, mutual authentication, confidentiality, integrity, and privacy protection remain unchanged compared to existing technologies, the 5G use cases and the associated new technical concepts do have an important impact on the security architecture.

The following list summarizes some of the key aspects of the up-to-date 5G security architecture:

- New and already existing service and device deployment scenarios for mIoT
- Energy efficiency, especially for battery driven devices
- Compatibility with ultra-low latency requirements
- New technical concepts such as network slicing and virtualization
- Enhanced user data integrity

- Enhanced user privacy
- Different types of access networks ranging from licensed to unlicensed spectrums
- Public vs. private networks, i.e. open vs. closed networks

As a general note, it is important that security is integrated and added by design into the native environment instead of later retrofitting into the networks and/or devices.

The following sections present more specific security challenges for the new 5G functionalities.

8.2.3 Network Slicing

According to [3], the slicing of a public land mobile network (PLMN) is not visible to the UE at the radio interface and it is assumed that the isolation (i.e. the slice) does not extend into the UE. Based on this assumption, the Next Generation Mobile Network (NGMN) has carried out a security analysis and issued a set of security recommendations addressing network slicing in [4, 5]. The respective issues to be considered include:

- Protection of the interfaces of network slices and functions for internetwork slice communication
- Prevention of impersonation attacks against the network slice selection function
- Prevention of impersonation attacks against the network slice instance
- Assurance of adequate mechanisms to provide sufficient protection if different security protocols or policies are applied in different network slices
- Protection against DoS through exhausting resources that are common to multiple slices within one slice
- Prevention of side channel attacks and timing attacks across slices when several slices share the same underlying hardware
- Consideration of hybrid deployment models, i.e. a mixture of virtualized and nonvirtualized network functions
- Provisioning of security mechanisms within the network or potentially within the UE to enable sealing between slices when the UE is attached to several slices

The network slicing concept requires the already familiar base security as for the UE when it performs the security access procedure with the slice. In addition, there are new points of view to be considered such as the guaranteed isolation between the slices. This means that if under any circumstances any single slice is compromised, it should not allow access to any other slices. The proper isolation ensures the protection for the integrity and confidentiality. Furthermore, it is of utmost importance that the resources, whether they are related to the network or slice instance, are not compromised between the slices if any of the slices experiences issues with the attacks.

In 5G, the UE may use variety of resources accessing within the 5G as well as via non-3GPP networks, to use variety of parallel services. If a single network slice data is tampered, an unauthorized UE may potentially aim to utilize the obtained data for establishing connection with the network slice for accessing and utilizing the resources, which need to be thus protected by the network operator.

One of the benefits of the network slicing is the possibility of defining different levels of security per slice, which refers to the authentication and authorization of the accessing users, and utilization of certain set of security functions to hosting applications.

This tailored security approach optimizes the network performance for different use cases; as an example, critical communications (CriC) use case may require stronger protection than mIoT use case. In the balancing of the security level, it should be noted that heavier protection also requires more resources and can consume more time, which, in turn, may negatively affect the latency. So, this is again one of the many network planning optimization tasks of the operator.

8.2.4 Network Virtualization

With regards to the assurance of the security of network virtualization, we need to consider the following aspects [6]:

- Virtualization must not cause new security threats, e.g. caused by vulnerabilities of the hypervisor.
- (Common) network functions and hardware resources such as storage and networking are connected throughout different virtual NFs.
- There is interconnectivity between network functions virtualization (NFV) components.
- Isolation of virtual network functions (VNFs) are used over the NFV infrastructure.
- Performance – it is likely that VNFs may have less per-instance capacity compared to physical NFs. Therefore, mechanisms are needed that split the workload between multiple instances of the VNF.

8.2.5 Edge Computing

As MEC is based on the principle of virtualization, the same security considerations as for NFV apply [7]. In addition, the NGMN has identified several key issues in [8]. Some of the respective security issues are described as follows:

- Routing significant data directly between the UE and the network edge without involving the core network and, in roaming, without reaching the home network implies that the charging records transmitted by the edge component have to be reliable and secured. Considering that the edges of the network are more vulnerable to attacks the risk of billing errors is potentially significant.
- Edge computing applications run on the same hardware as NFs do. As edge computing applications may not be owned and controlled by the operator, a relationship between those entities (operator and third-party application provider) must be established that guarantees that the third-party application does not harm the operator NFs.
- It is intended that applications influence the performance of the network automatically. It means that in case the applications require more bandwidth, the NF at the edge is configured automatically so that the demanding application gets more bandwidth. This may lead to reduced performance and throughput for other applications if the demanding application consumes too much bandwidth either accidentally or maliciously.
- Delivering content directly from the edge to the UE implies that replicas of the content are available within the edges in a cashed form. Accessing this content means that Internet Protocol (IP) communication terminates and that IP protocols have their endpoints in the edge.

- Storage of sensitive security assets at the edge need to be protected to avoid compromising them. In addition, those assets need to be protected when being exchanged between the network core and the mobile edge.

  > If latency targets are too aggressive, this can pose quite severe constraints on security mechanisms. The benefits and implications of very low latency need to be carefully weighed against each other; moreover, operators should not overlook security to reach latencies that might not be absolutely needed to start with [8].

It is stated that if the latency requirement is less than 1 ms, the core network node can be at most 150 km away from the UE (assuming the speed of light in glass fiber) and may thus be applicable to private networks (e.g. factory use case). Low latency may also include authentication in the beginning of the attachment or handover procedures. In case of roaming, this may have to be done entirely within the visited network as a round trip to the home network may take too long. In this case, solutions such as "data-efficient re-keying" as described in Annex B2 of [9] may be appropriate.

8.2.6 Open Source

Introducing standard IP means that network infrastructure that has previously been protected through "security by obscurity" and proprietary mechanisms is now subject to a much larger attack community. It is expected that this also opens new attack vectors. All successful attacks on IPs now also automatically affect the 5G infrastructure [10, 11].

8.2.7 Other 5G Safety Considerations

In addition to the above-mentioned technologies, various other new trends must be considered, as they may play an important role when defining the appropriate security technology for 5G. The following list summarizes a few examples:

- *Quantum-safe computing.* As 5G is supposed to be used during the next 20 and more years, quantum computing may develop into the next levels, jeopardizing all the previous security measures. Specifically, quantum computing may allow breaking asymmetric crypto algorithms that are considered future-proofed still up today. New asymmetric algorithms may be needed to overcome this threat. In addition, symmetric algorithms are regarded as quantum-safe. However, it may be required to extend the length of the keys.
- *Physical layer security.* Today, security takes place above the physical layers. For a second phase of 5G (e.g. when using the new spectrum) physical layer security may have to be considered. Through this means, low latency and high performance could be achieved when those mechanisms are implemented in the physical layer. Some concepts exist and are being discussed in the research community.
- *mmWave.* A technology that is required to provide a new spectrum in the higher frequency bands, above 6 GHz. In addition to the reallocation of existing spectrum to 5G (e.g. 700 MHz for mIoT and 3.4–3.8 GHz for enhanced Mobile Broadband [eMBB] and vehicle-to-everything [V2X]), new spectrum such as 28 GHz in the United States and Korea is allocated and required to achieve indoor penetration, high bandwidth,

and low latency. Europe prefers 25 GHz and/or 32 GHz. It is important to harmonize the frequencies globally and to make sure synergies can be achieved by selecting frequencies so that they can be filtered and adapted (possible within a range of 4–5 GHz). This is important for devices but also for infrastructure components.

In addition, data offloading will continue to evolve, introducing super small cells that use nonlicensed band and enable seamless handover and roaming between nonlicensed and licensed bands.

8.3 Development

8.3.1 LTE Evolution

Once the standards are ready, the deployment 5G will take time. It requires a migration plan and proven interworking with other technologies, as well as backward compatibility with legacy mobile technologies. The following key trends on this path are identified as these will also impact the security: Long-Term Evolution (LTE), low-power wide area (LPWA) IoT, and migration and interworking.

Many operators are currently investing in LTE. It is likely that 4G-LTE will continue to grow and advance to address some of the use cases that are listed for 5G. It seems to be common understanding of the industry that the LTE-Advanced paves the way for 5G and that early 5G deployments may in fact be based on the 4G LTE-Advanced.

8.3.2 NB-IoT vs. Low-Power Wide Area

The proprietary LPWA systems such as LoRa and Sigfox are seen by the MNOs as bridging technologies toward 3GPP-standardized NB-IoT. Before the 5G mIoT arrives, technical and security challenges need to be solved including the deployment, energy, and cost efficiency as well as group operation. It is expected that 5G mIoT will replace NB-IoT later as 5G achieves additional cost advantages due to virtualization and network slicing. It is worth noting that IoT is a very wide concept with diverse use cases and that especially the low-cost use cases are driving the discussions around the requirements, including those of the security.

8.3.3 Ramping Up 5G

5G will be deployed in mature markets first. First trials and launches are likely to happen in Asia for the Olympic games in 2018 and 2020. The ramp-up may be driven by some specific use cases and in some specific regions. There will be a requirement to interwork with legacy systems that have defined security schemes and mechanisms. This includes authentication and key derivation as defined in the LTE based on the USIM functionality.

It is likely that not all new technology concepts as described above will be ready from the beginning. As an example, the network slicing may be introduced at a later phase of 5G to speed up the initial deployments. From a security point of view, it means that the MNOs will not be able to provide the level of isolation as can be achieved with network slicing and NFV.

Therefore, MNOs may decide to launch the 5G network, say, for a dedicated building block. This building block may support, e.g. the eMBB and some limited set of other features. The MNO may later provide updates to their network infrastructure and add gradually features and virtualized functions as the network matures. The installed base of devices may need to support that ramp-up strategy and allow the possibility to flexibly manage security configurations and mechanisms on the device.

8.4 Security Implications in 5G Environments and Use Cases

To facilitate and simplify the discussion, the 3GPP has split 5G into four building blocks. They are the basis for the definition of requirements, use case groups and common characteristics. The 3GPP building blocks are:

1. *Network operations* (NEO), defining the NEO use cases and building basis for the other building blocks
2. *Enhanced Mobile Broadband* (eMBB), defining high bandwidth and user mobility as well as broadband everywhere use cases
3. *Massive Internet of Things* (mIoT), addressing the low-power, wide-area range of use cases
4. *Critical communications* (CriC) for low latency and high reliability use cases
   Please note that V2X communication could be considered as a fifth building block or as a combination of the other blocks.

Figure 8.3 clarifies how the CriC, mIoT, and eMBB are based on the NEO. It also depicts that dedicated use cases such as virtual reality may be associated to a building block, whereas other use cases cannot be associated with only one building block (e.g. V2X) but, rather, belong to several building blocks. For more background information on use cases and building blocks, please refer to [2, 11].

8.4.1 Network Operations

The main target of the MNOs is to build the 5G networks in a flexible manner to support diverse scenario demands, e.g. by introducing slicing of the network for various market

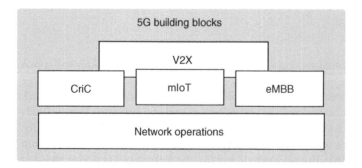

Figure 8.3 5G building blocks as defined by 3GPP are critical communications (CriC), massive IoT (mIoT), enhanced mobile broadband, and their respective network operations (NEO). Sometimes also vehicle-to-everything can be interpreted to form another 5G building block.

segments and verticals. This building block covers use cases for enabling operators to support an elastic and scalable network to optimize use of mobility management for diverse scenarios and to support efficient content delivery. The key aspects for the NEO building block are flexibility, bearer-independency, and scalability.

It also addresses access related use cases including selection of the most appropriate access for user traffic, coexistence with legacy systems and the migration of services from early generations. In addition, the exposing network capabilities to third parties are handled within this building block. The following security requirements need to be covered:

- Confidentiality and integrity protection of voice, data, and signaling
- Authorization, confidentiality, and integrity protection between network elements and networks
- Authorization, confidentiality, and integrity protection for next generation services
- Authorization for users, devices, and networks
- Extensible systems for new algorithms and procedures to mitigate risks e.g. coming from quantum computing
- Prevention against DoS and signaling attacks
- Privacy protection, e.g. by means of pseudonyms or temporary identifiers and by protection user location information
- Support emergency cases i.e. granting temporary access to the network based on operator policies
- Device theft prevention: supporting a secure mechanism to disable or re-enable a stolen device; protecting device identifiers when being stored

With regards to the new technologies and concepts introduced with 5G this requires the security solution to consider the following topics:

- *Service specific security as realized by a network slice must be supported.* Those security configurations are well isolated within the network by the means of those slices. Slice-specific security includes the capability to leave it up to third parties to configure the security of that slice, e.g. by suitable application programming interfaces (APIs) to a certain extent (as defined by the network operator).
- *The 5G system shall support a secure mechanism to collect system information while ensuring end-user and application privacy.* For example, application level information such as application usage information is not to be related to an individual application user identity or subscriber identity and UE level information, such as UE location is not to be related to an individual subscriber identity.
- *LI requirements need to be fulfilled.* That also covers content that may be cached e.g. at the edge of the network.
- *Many ways to access the network shall be supported.* This includes the case where a device accesses the 5G network through a different radio access technology (Wi-Fi) and authenticates using its 5G credentials.

Further requirements and statements on security are given for each building block with an emphasize on the most important ones. In addition, please refer to the requirements defined by 3GPP SA1, as given in their technical report on NEO in [12].

8.4.2 Enhanced Mobile Broadband

The eMBB is designed to protect high value content and connections. This building block covers use cases that require high data rates, high density (e.g. to serve capacity peaks for the football stadium events), and high user mobility (including airplane communications).

It identifies key scenarios from which the eMBB primary data rate requirements for peak, experienced, downlink, uplink, etc. can be derived. It includes scenarios for the transport of high volume of data traffic per area (traffic density) or transport of data for a high number of connections (devices density or connection density).

The eMBB is a natural evolution of LTE and LTE-Advanced. It is characterized by voice and large bandwidth use cases that involve both high value content and high value connectivity such as video streaming, virtual reality, or augmented reality. The following security capabilities need to be considered:

- *Content protection.* The content is valuable and must be sent to authorized and authenticated users only. The content may also need to be protected by encrypting it both on the network layer (via encryption and integrity protection on the radio interface) and the service layer (DRM service). In addition, content may be broadcasted to several devices. In those cases, the security applied must be able to support broadcasting scenarios and must authenticate a number of devices when sending the content to them in protected form.
- *As users are moving indoor, outdoor, or between indoor and outdoor, the authentication mechanism need to support a seamless user experience.* This means the connection must remain stable with no interruption even when moving from Wi-Fi (indoor) to the 5G network (outdoor), thus not requiring a reestablishment of the session because of reauthentication. To support those scenarios, the authentication end point may need to reside in the core of the 3GPP network for both indoor and outdoor usage and handover must be supported between the two networks. In addition, a separate authentication method may be needed to establish initial connection to the Wi-Fi network (indoor).
- *Privacy is important.* Users do not want others to determine which data they consumed when and where.
- Strong mutual authentication and user identification is required. This enables charging/billing.
- *Fixed and mobile convergence must be considered.* Some fixed-line services may be replaced by mobile services to address the last mile to the customer. In that case, the overall security over fixed broadband shall be the same as for the 5G mobile broadband access link.

8.4.3 Massive Internet of Things

The mIoT building block is designed to be flexible and efficient. This building block addresses use cases that involve large numbers of devices with non-time-critical data transfer. Devices are either simple or complex and need to consider efficiency of security and device configuration. It includes smart wearables (personal area network) that are characterized by low complexity, high battery life, high reliability, and sometimes high data rates. It also includes sensor networks that are characterized by a very high density

Table 8.1 Needs for massive IoT protection.

Key area	Needs for protection
Service deployment	• OEMs are required to produce large volumes of low cost devices efficiently. • Provisioning of devices in the field need to be efficient, even when there no initial subscription on the IoT device. • Device life-cycle management includes need for dynamic subscription management in addition to statically provisioned subscriptions.
Network and service access	• Devices accessing 3GPP RAT and non-3GPP RAT connect to the same network core. • Direct or indirect (via relay device) connection to the network • Roaming and nonroaming cases
Efficiency	• Optimize battery consumption • Reduced signaling effort and network overhead for security
Group operation	• Data to be transmitted to a group of devices • Provision and authenticate a bulk of devices • Lightweight mechanism for device configuration (non IMS enabled devices)

of devices, very low complexity with very long battery life, e.g. for smart services in urban, suburban, or even rural areas. Nevertheless, even though the IoT field of use cases is very diverse and wide the discussion on security requirements is mainly driven by low-cost use cases.

Table 8.1 summarizes some of the key areas that need to be considered in the designing of mIoT services.

It is generally thought that there will be a massive number of connected devices soon, all of them submitting sporadic, small-scale data. These devices may reside in the field for a long lifespan of up to 10–15 years or more. From a security point of view, the valuable assets that need to be protected are data and identifiers. However, even if the value of the network subscription may be rather low, the number of devices and the risk of misusing those devices to overload or spam the network need to be considered.

The following security capabilities in mIoT security need to be considered:

- *Data represents the asset in mIoT*. Therefore, it is necessary to prevent unauthorized manipulation of that data and to thus ensure data integrity. A separation of the application security from the network access security is in particular relevant in IoT as in many IoT scenario the value of the application data may be more significant compared to the value of the single connectivity.
- *Data need to be confidentiality-protected*. This is for privacy reasons and to prevent misuse.
- *For battery-powered devices, security protocols and mechanism need to be energy-efficient*. They must also provide high security while optimizing the overhead, e.g. by reducing reauthentication intervals.
- *Security expenses must be kept to a minimum*. As the cost of such a considerable number of devices need to be low, the extra expense resulting from the respective security mechanisms need to be reduced to an absolute minimum.

- *mIoT devices may access the network either directly or indirectly through licensed or unlicensed bands.* It is essential to provide the authentication endpoint in the core of the network to make sure the appropriate credentials are used to access the service. It must not be possible to misuse lower security mechanisms of networks such as Bluetooth, Wi-Fi, or others to bid down or infringe the security of the service and the overall network.
- *Data manipulation must be prevented, and privacy provided even when the mIoT device is connected via a relay device, i.e. having indirect connection.* Additional authentication and authorization mechanisms may need to be supported to access several types of connectivity.
- *mIoT devices must determine if a relay device is authorized to act in such a role to the 3GPP network.*
- *As some device categories will be produced in large volumes, pre-provisioning with connectivity or service credentials may not be feasible without adding cost or complexity.* Therefore, secure provisioning of credentials of devices in the field is required.
- *Device-to-device communication and group aspects need to be considered as well.* This includes appropriate and efficient authentication mechanisms for a group of devices as well as sending the same data to a group of devices.

8.4.4 Critical Communications

CriC covers all use cases that support low end-to-end latency ranging from 1 ms up to 10 ms even in the highest-mobility scenarios. Its security must be by default both fast and reliable. The 3GPP system shall support ultra-high reliability (99.999% or higher) and high uplink data rate (tens of Mb/s per device in a dense environment). Also, the system needs to support dynamic resource utilization in the cloud and at the network edge (compute, storage, network, and radio) for a given UE and shall support higher accuracy location capability.

CriC requires robust security. Many use cases, such as autonomous driving or remote surgery, involve human well being, including the lives. If security is weak or does not fulfill the high demands of that use case, this human life is at stake! Other use cases require tactile Internet experience to remote control robots or drones. In those cases, insufficient security or delays introduced by security may cause significant loss of value.

The following security capabilities need to be considered in particular:

- *Data confidentiality.* Data must be protected in transit and at rest both in the network and the device to protect privacy and to avoid unauthorized access to this data.
- *Data integrity.* It is of utmost importance to make sure that data cannot be manipulated. Imagine a command send by the doctor to the remote surgery robot is modified – this will immediately have severe negative impact on the human being treated. The origin and authenticity of the data need to be verifiable.
- *Strong mutual authentication.* This can deal with the low latency requirements of the use cases. Authentication protocols and frequency of reauthentication need to be realized in a way that latency requirements can be fulfilled. However, as security in this building block is particularly important, we must not compromise the level of security to benefit the latency. In addition, fast algorithms and optimized protocols may have to be implemented to realize the low latency requirements. In addition, storing and processing the identifiers/identities of the device and the subscriber in a secure way is important. Those identifiers are used for authentication and other security purposes, and therefore both need to be kept separated and need to be protected well.

- *Security mechanisms must be highly reliable.* Strong and formally verified protocols as well as algorithms and processing must ensure that implemented security mechanisms are reliable and do not negatively impact the overall system reliability. In addition, security certifications may be appropriate to ensure that the implemented security levels fulfill the high demands.
- *For closed networks operated by factories or enterprises, third-party authentication schemes must be supported using identifiers and credentials provided and managed by that third party.* In addition to open/public networks 5G will be deployed in closed scenarios either by allocating a dedicated network slice to that, e.g. factory or by setting up a 5G network that is operated either by the factory itself, by the MNO or by the network infrastructure provider. The third party may want to use their implemented ID management system (e.g. based on certificates) to grand access to their network and company services. For this purpose, diverse types of credentials and alternative authentication mechanisms need to be supported.
- *In some cases, some devices must have prioritized access to the network over other devices.* For this purpose, it is essential that certain CriC devices are authorized to have prioritized access and that no other device can manipulate that priority.

8.5 5G Security Layers

In addition to the different security configurations that may be introduced by the network operators by the means of network slices, the security requirements of other stakeholders need to be considered. Those stakeholders include the consumer, the application provider, the service provider, and the original equipment manufacturer (OEM).

In order to reflect and address the requirements of those stakeholders and in order to separate them from the MNOs requirements during the discussion and definition of security solutions, SIMalliance has introduced a security layer concept in their technical white paper [13] as given in Figure 8.4. A detailed description of those layers can be found in that white paper and is not repeated here.

However, please note that this figure does not address the following aspects:

- The device ID and key used for device identification may reside within the SE.
- The SE could be used to also check the integrity of the device, e.g. by providing Trusted Platform Module (TPM) functionality on SE.
- There could be several tunnels to different applications.
- There could be multiple tunnels and keys showing that there could be multiple service providers.
- Service layer integrity protection could be combined with identification and authentication.
- The network layer is split into user and control plane where user plane (UP) integrity and encryption reach from the device to the service provider using a different session key. Note that Universal Mobile Telecommunications System (UMTS) and LTE do not include integrity protection of the UP data.
- Some key derivation could be provided by the SE and the network entity to derive further, higher layer key material and security based on the operators' key.

It should also be noted that in some cases, the radio interface security may be applied depending on higher layer security. In this case, the radio interface need to know about

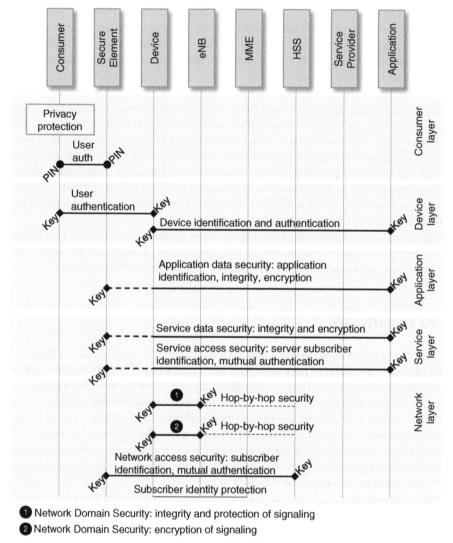

Figure 8.4 Security layers as interpreted from SIMalliance.

the security available in the higher layers. Nevertheless, how this can be done in practice is not defined yet.

8.6 Device Security

8.6.1 Differentiating the Devices

As discussed in this chapter, 5G devices need to support several and diverging use cases. Depending on those use cases and the capabilities, the device can be associated with a device class based on its characteristics including spectrum (5G only or dual 4G/5G, single/multiband), number of applications, type of network (licensed or unlicensed, public, closed, fixed, wireless), type of access (direct or indirect), and type of credential (5G or non-5G).

8.6.2 Spectrum

Most devices are going to support multiple bands to support backward compatibility and roaming in countries, allocating a different spectrum than the home country. This means they will have to support several network authentication applications to access the respective network. However, in some cases (e.g. factory use case, closed network) there may be devices that are only used within a dedicated region and therefore are required to only support 5G.

Devices supporting the new spectrum (e.g. 28 GHz) will most likely take a couple more years, as chipsets supporting those frequencies need to be developed and need to be small and cost efficient enough to be included in a wide range of devices such as smart phones.

8.6.3 Number of Applications

Certain simple IoT devices and sensors may have one specific purpose and will consequently be provisioned with one specific application. It is not foreseen for those devices that the application will change or be replaced or another application will be added. This means that throughout the lifetime of the sensor (which could be up to 15 years), the application will not change.

This may, however, not be true for the security configuration or keys on those devices. In case of a security incident, it may need to be required to replace the key or to change the security of the device as this may still be more cost efficient compared to replacing the device.

Other devices such as smart phones or vehicles will have to support many different applications at the same time. A vehicle may have to support autonomous driving as well as video streaming and data collection, thus representing use cases of all different building blocks as described above. Those devices need to be managed frequently to support new applications. In addition, those applications or network slices may support different security configurations that have to be supported by the device.

8.6.4 Type of Network

We currently distinguish licensed bands that can provide a certain quality of service and service level agreements and unlicensed bands that, by contrast, provide access at lower cost. There are devices that support either access to licensed band or to both licensed and unlicensed band.

Other devices do only support access to unlicensed band. A network may be public, i.e. widely deployed and accessible by many subscribers or closed (e.g. factory). Closed networks may be operated by third parties either independently or with the support of either the MNO or the network infrastructure provider. Also, fixed and wireless networks need to be covered within 5G.

8.6.5   Type of Access

Depending on the type of network, the device may be connected to the 5G network directly or via a relay device. In case the device does not support direct access, it may still hold 5G network credentials to authenticate to the 5G network when accessing it via the relay device.

8.6.6   Type of Credentials

Devices may support different types of credentials, depending on the type of access and type of network they support. In addition, devices that are used within public or closed networks and may need to support different types of credentials for each purpose.

There may be devices that hold several completely independent credentials or multiple credentials of the same type or a combination of both. Other devices that are bound to one specific type of network or access and that are bound to a specific use case may only support one type of credential.

8.6.7   Devices for Network Security

Devices are used to access and consume services. This can only be achieved if the devices are provisioned with a subscription that enables connection to the network. To connect, they need to be able to perform mutual authentication with the network. Today, this authentication is done based on the USIM residing on the UICC. In addition, the UICC is used to derive additional keys for integrity and confidentiality protection on the network radio interface. Hence, the devices today use 3GPP security mechanism to protect the network layer.

In 5G, being the network of networks, devices need to access different types of networks (see "type of network" and "type of access" above). That means that multiple different authentication mechanisms are required to support that. In addition to accessing the network using the network subscription, it is envisioned that the device has its own credentials (keys and identifiers) to enable authentication of the device independently of the authentication of the network subscription (represented by credentials independent of the device).

Let's start with devices accessing closed networks. The assumption is that a factory or enterprise is operating its own 5G network. To grant employees access to the corporate network the factory may want to use its existing company ID system and thus replace the company ID card by the mobile device both for physical access and for access to in-company services and network resources. To authenticate the user/device and the network, a certificate-based company ID system is used as authentication entity within and connected to the core of the closed 5G network. The device in this case needs to support the certificate-based authentication and needs to be provisioned with the respective credentials and certificates.

At the same time, the device should also be usable when the employee leaves the factory building and when the closed network does not provide coverage anymore. In this case, a handover to the public network of the network operator should take place based on credentials and authentication mechanisms owned and controlled by the network operator, as shown in Figure 8.5.

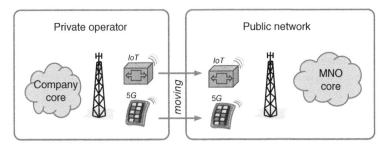

Figure 8.5 Handover between closed and public network.

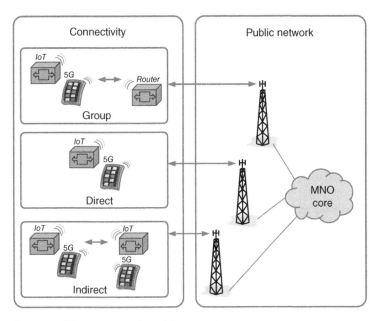

Figure 8.6 Ways to access the public network.

Access to public network can be achieved in several ways. Figure 8.6 shows how single devices or a group of devices can access the network. This can be done through a gateway, directly if the device has cellular capabilities or indirectly using a relay device in case the connecting device does not have cellular capabilities. In addition, devices are supposed to connect to other devices without any connectivity to the public network.

The following security requirements can be derived from those different scenarios:

Group access. Authentication of groups of devices is required, in addition still each device may need to have its individual identifier to allow targeted management operations on that device. Group management of credentials and update of security capabilities is needed. Mechanisms defined for this purpose need to limit the impact on the network and signaling resources.

Direct access. Authenticated and authorized access to the network is needed. This is done based on network access credentials. There is a need to be able to manage those credentials.

Indirect access. The device connecting to the core need to be provisioned with network access credentials, however, as it does not have its own cellular connectivity it is using a relay device to transmit and authenticate with the core using its own network access credentials. Those are tunneled to the serving network via the relaying device. That relaying device may have its own credentials to establish a direct connection with the serving network.

Specific security requirements include the following:

- Group authentication and security management using group identifiers and keys
- Updating and managing security settings and configurations of groups, as well as of gateways that may reside in between
- Verifying that the relay device is authorized to act as a relay as well as enabling the network to authorize relay devices to act as a relay
- Tunneling credentials and performing authentication of relayed devices via the relaying device

8.6.8 Device-to-Service and Device-to-Application Security

Today, several features are defined for the USIM that could be used to provide security also on the higher layers. Those features include the Generic Bootstrapping Architecture (GBA), as defined in [14], that works to derive additional key material from the USIM. That key material can be used to set up a unique and time-limited session key used to set up a secure TLS (Transport Layer Security) tunnel with the application server. Extensible Authentication Protocol, and Authentication and Key Agreement (EAP-AKA) as well as EAP-AKA' are used to authenticate to non-3GPP networks such as Wi-Fi.

Also, several other already-specified UICC mechanisms may be used to provide security on the upper layers. GSMA Association (GSMA) summarizes these USIM features with a focus on enhancing authentication to IoT services in [15].

In current generations, the network operator is the owner of the UICC. If the network operator decides to offer additional services or applications, thus acts as service provider or application provider, the additional mechanisms could easily be enabled and used. However, it turns out that those mechanisms did not result in commercial and market success.

In 5G, the ownership model of the UICC continue to change. With technologies such as embedded UICC (embedded subscriber identity module, eSIM), the ownership model of the security component already changes. In this scenario the OEM owns the security component that opens the possibility to address the security needs of both the OEM and the MNO. The deeper the integration into the device is done ranging from a pure hardware integration to a full software integration, making the security component accessible by specific APIs, the easier it will be to open this component to additional stakeholders, including application providers.

To access services, a service subscription can be provisioned and stored securely on the device. Applications may leverage specific security functions provided by the security entity residing within the device or they may continue to implement their own security schemes e.g. using obfuscation, whitebox cryptography or other SW protection mechanisms.

Specific security requirements include the following:

- A generic security platform within the device must provide SW interfaces and APIs to applications.
- Security platform shall be able to handle security configurations of multiple different and completely independent stakeholders.
- Device shall be able to hold service subscriptions in addition to network access subscriptions and use those to authenticate to the service.

8.6.9 Device-to-Device Security

In CriC, 5G devices will have to be able to directly communicate with other devices – as is the case with push-to-talk today. Police or fire brigades require to use radio in emergency areas even when there is no network coverage. The respective security requirements include:

- Mutual authentication of a single or a group of devices – this involves keys need that to be negotiated and agreed.
- Data must be encrypted and integrity protected.

8.6.10 Device-to-User Security

The main security requirement to be fulfilled is that only authorized users shall be able to access the device. This can be achieved by verifying a PIN or biometric characteristic of the user. In addition, theft prevention should be considered, i.e. only authorized entities shall be able to disable or re-enable stolen devices.

Also, users may want to store and process private and confidential data on the device (e.g. photos) – applications leveraging the secure entity within the device could be used for this purpose.

8.6.11 Summary of the Device Security Requirements

Figure 8.7 summarizes the high-level security requirements that the 5G device must fulfill.

5G device security must be:

- Supportive of multiple security layers and thus supportive of multiple different and independent stakeholders
- Future-proof and flexible to allow reconfiguration and management of security settings and credentials and to support further evolution of 5G
- Fast and powerful to comply with low latency requirements
- Certifiable and highly reliable to fulfill the high demands of the CriC use cases
- Tamper-resistant to security store sensitive data, security configurations and credentials
- Backward-compatible to support migration toward 5G and to be able to support legacy systems
- Managed remotely to enable the flexibility requirements above
- Energy and cost efficient to be suitable for a wide range of devices, including LPWA IoT devices

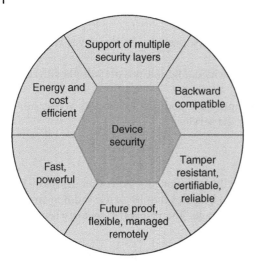

Figure 8.7 High-level security requirements.

8.7 Security Between Network Entities

8.7.1 Protection of the Mobile Edge

With MEC and virtualization, network functions and content are moving closer to the consumer. That means that NFs and content are replicated many times and are available in environments that are less protected than the network core (Figure 8.8).

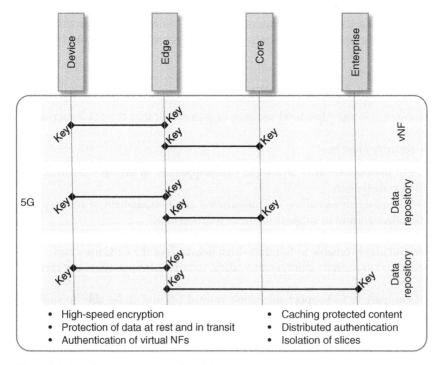

Figure 8.8 Security between network entities.

Edges need to retrieve an *instance* of the network function from the core. To make content available to the user with reduced latency, the edges also need to cache content provided by the third-party content provider.

To make sure the edge is authorized to receive the NF instance or the content, authentication between the involved network entities need to take place. In addition, the data being exchanged between those entities need to be protected at rest and in transit. This can be achieved by applying either software- or hardware-based security mechanisms.

Virtualization is a main concept in 5G so the assumption is that the security mechanisms in the edges will also be based on virtualization technologies involving hypervisors and isolation. For performance and enhanced security reasons, however, it may make sense to integrate hardware security modules (HSMs), in network entities that are exposed to and accessible by attackers. HSMs also provide secure handling of credentials and can be used to store and derive authentications vectors used to authenticate access to the network.

8.7.2   Authentication Framework

As already described, there will be different ways to authenticate access the network. To support fixed and wireless networks as well as licensed and unlicensed, public and closed networks, and several different network slices, a flexible and scalable authentication framework is required both on the device and the network infrastructure. This implies that several authentication mechanisms need to be supported on the device as well as by the respective network entities (Figure 8.9).

8.8   Security Opportunities for Stakeholders

8.8.1   Evolution of the SIM – Scalable Device Security

Several use cases require strong protection mechanisms and a security solution that is both efficient and reliable. Wide range of products or technologies could be used to provide additional security within the 5G device. This includes SW security solutions such as TAK, Whitebox crypto, code protection, and hardening, which can be applied on the application layer. It also includes solutions that are based on hardware-enabled

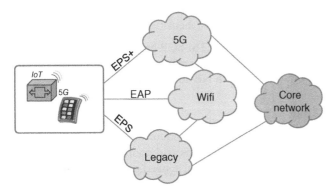

Figure 8.9 Examples of the 5G connectivity.

security technologies that are deeply integrated into the chipset such as ARM TrustZone or Intel SGX and reach to dedicated security hardware being either implemented as a distinct chip such as UICC or eUICC with own resources or deeply integrated into the chipset such as for the iUICC. For more details on the various types of xUICCs and integration levels, please refer to [16].

To achieve a balanced relation between security level versus the bill of material of the device and to be able to fulfill new technical requirements, such as performance and power efficiency, the SIM need to evolve to become an "evolvedSIM." The evolvedSIM need to be understood as an advanced secure platform:

- *It has an optimized form factor for its specific usage.* The assumption is that removable plastic SIM cards will only be suitable for the transition phase toward 5G and very specific use cases that may require this feature. The form factors will have to be optimized to facilitate an efficient and cost-effective integration into the devices. The evolvedSIM may be very deeply integrated into the device. In the extreme, the evolvedSIM will need to have no dedicated form factor at all and will be implemented as packaged UICC, vUICC, or iUICC, as described in [16].
- *It processes multiple security contexts simultaneously.* The evolvedSIM must be able to process different applications at different layers and for different stakeholders. This reaches from supporting different network authentication applications for the different slices or networks provided to application layer security and device security features, such as secure storage of device identifiers and verification of device integrity.
- *It provides appropriate interfaces that enable an integration into devices in an efficient way.* Those interfaces need to be powerful enough to cope with low latency requirements as well as with high data rates and increased security demands of use cases such as CriC. In addition, interfaces that are more relevant in the embedded world need to be considered that enable reduction of integration and development effort of device manufacturers. For technologies such as vUICC and iUICC, new interfaces may need to be introduced allowing external hardware components (external to the processor hosting the UICC) or software to access the UICC on that chip. This requires respective APIs as well as the usage of standard protocols such as Serial Peripheral Interface (SPI) or Inter-Integrated Circuit (I^2C).
- Offers enhanced security capabilities to several stakeholders. As just described both APIs allowing applications to leverage the evolvedSIM and a clear separation of UICC applications or security contexts need to be ensured in order to allow other stakeholders to access and leverage the enhanced and smart security platform.

As an example, from the standardization field, ETSI SCP has launched a work item dealing with some of these aspects. Also, some major chipset vendors are working on vUICC and iUICC concepts as a solution for both NB-IoT and 5G.

8.8.2 Protecting the Mobile Edge

The new concepts such as virtualization and MEC demand for increased security measures and solutions. The edges hold network functions and content, so they need to provide adequate protection for cached content. They also need protect the authentication vectors (AV) and other sensitive data.

On the other hand, the network core or the enterprise servers providing either the network function instances or the content to the edge must ensure that the information and data are only sent to authorized and genuine edges. For this purpose, mutual authentication at network function level as well as on network entity level is required.

8.8.3 Providing Virtualized Network Components

Virtualization will start with network functions residing in the core. It will begin with basic functions and extend step by step. Based on this trend, it may be possible to adjust the current solutions of the security providers.

8.8.4 Solutions for Closed Networks

Next to public networks, a huge number of closed networks for factory or enterprise use cases is expected. Network infrastructure providers put a lot of effort into creating this market and to increasing their customer base from MNOs to such operators of closed networks. Those closed networks will have to deal with different authentication mechanisms and technologies. Those mechanisms will have to be implemented within both the device and the closed network and will in many cases be based on an ID system (e.g. certificate-based system) that is either already implemented in that enterprise or that will be established. Implementation of respective ID management solutions and the management of the credentials with tailored ID management systems or hosted services are thus important.

8.9 5G Security Architecture for 3GPP Networks

This section summarizes the relevant information available up-to-date in standardization, related to network security functions and elements, including the updated procedures and security processes from previous systems and aspects within the network and in interoperable modes; new methods for security monitoring, enhanced LI. This section is largely referring to the 3GPP TS 33.501, as defined in Release 15, version 15.0.0, which was published in March 2018 [17].

8.9.1 Network Security Functions and Elements

The key specifications for the 3GPP security architecture, security domains, and procedures are TS 33.501 and 33.401. This section presents the most essential aspects found in these sources.

The 3GPP 5G system consists of evolved physical radio and core networks and their logical functions. The security architecture can be presented as *security domains* that consist of *application stratum, home and serving stratum*, and *transport stratum*. Figure 8.10 outlines the functional elements within these stratums and their security relations.

Figure 8.11 depicts the 5G network functions that are related to the security. This example is applicable to the nonroaming scenario. In addition to these security interfaces, 3GPP also defines a Security Edge Protection Proxy (SEPP), which is meant to protect the messages of the *N32* interface. The security interfaces shown in Figure 8.10 are described in Table 8.2.

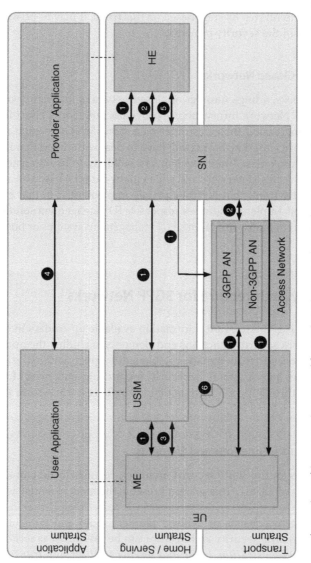

Figure 8.10 3GPP system's security architecture as interpreted from 3GPP TS 33.501. Please refer to Table 8.2 for the description of the interfaces.

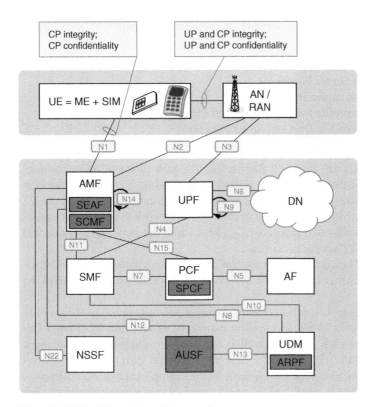

Figure 8.11 The 5G security architecture for nonroaming scenario.

Table 8.2 Description of the security architecture interfaces as presented in Figure 8.10.

Interface	Function
1	Network access security features for UE to authenticate and access services securely via network. These features protect the radio interface, whether it is 3GPP or non-3GPP radio access, and deliver security context SN to UE.
2	Network domain security features provide means for network nodes to securely exchange user data and signaling.
3	User domain security features secure the user access to ME.
4	Application domain security features provide means for applications of user and provider to exchange messages securely.
5	Service-based architecture (SBA) domain security features include network element registration, discovery, authorization security, and protection for the service-based interfaces.
6	Visibility and configurability of security features provides means to inform user if a security feature is in operation.

In the initial phase of the primary authentication, the 5G functional elements involved in the signaling between the UE and network are depicted in Figure 8.11, as stated in Ref. [17]:

- SEAF (Security Anchor Function)
- AUSF (Authentication Server Function)
- UDM (Unified Data Management)
- ARPF (Authentication credential Repository and Processing Function)
- SIDF (Subscription Identifier De-Concealing Function)

The SEAF is a security anchor that is collocated with the Access and Mobility Management Function (AMF) in 3GPP Release 15. The task of the SEAF is to form, because of the primary authentication, the unified anchor key K_{SEAF}, which is common for all the access scenarios. This key is applied to protect the communications by the UE and the serving network. The K_{SEAF} resides in the visited network in roaming scenario. Please note that there may be two separate K_{SEAF} keys when the same UE is connected to a 3GPP and a non-3GPP access networks.

The Security Context Management Function (SCMF) retrieves the key from the SEAF, which is used to derive further keys. The SCMF may be collocated with the SEAF in the same AMF.

The Security Policy Control Function (SPCF) provides policies related to the security of network functions such as AMF and session management function (SMF) and UE. The elements involved for each policy scenario is dictated by the application level input of the application function (AF). The SPCF may be collocated with the policy control function (PCF), or it can be a stand-alone element. Some examples of the tasks the policy may contribute to include the confidentiality and integrity protection algorithms, key lifetime and length, and the selection of the AUSF.

The AUSF replaces the Mobility Management Entity (MME) and Authentication, Authorization, and Accounting (AAA) of the 4G system. It terminates requests from the SEAF and interacts with the ARPF. The AUSF and the ARPF could be collocated to form as general Extensible Authentication Protocol (EAP) server for EAP-AKA and EAP-AKA', in which case the deployment needs to comply with the 3GPP TS 23.402.

The ARPF is collocated with the UDM element. It stores the long-term security credentials such as user's key K. It executes cryptographic algorithms based on the long-term security credentials, and it creates AV.

The SIDF is a service offered by the UDM network function of the home network of the subscriber. Its task is to de-conceal the Subscription Permanent Identifier (SUPI) from the Subscription Concealed Identifier (SUCI).

8.9.2 Key Hierarchy

The 5G system has a key hierarchy as depicted in Figure 8.12, defined by the 3GPP. The defined key lengths are by default 128 bits in the initial phase of the 5G networks, and the network interfaces shall be prepared to support key lengths of 256 bits in the future.

Table 8.3 summarizes the principle of the keys used in 5G system.

In addition to the keys shown in Figure 8.12 Table 8.3, there is a set of other intermediate keys. K_{gNB}, is one of these, derived by Mobile Equipment (ME) and 5G NodeB (gNB) in the derivation of horizontal or vertical key, using Key Derivation Function (KDF).

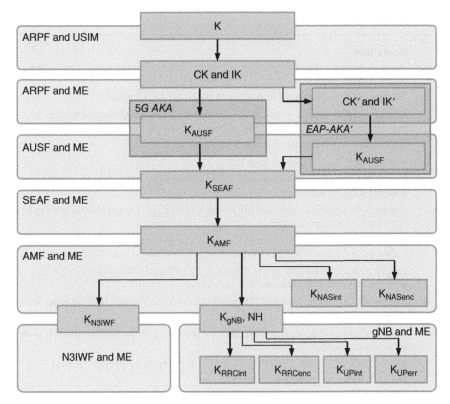

Figure 8.12 The 5G key hierarchy as defined by 3GPP in TS 33.501.

Another one is K′$_{AMF}$ which is derived by AMF and ME when the UE moves from one AMF to another during inter-AMF mobility, using KDF.

Also, for the case of dual connectivity, as described in Section 4.2 of this book, there is a key derived. This happens in such a way that when the master node (MN) establishes security context first time between slave node (SN) and the device (UE) for a access stratum (AS) security context, the MN forms an S-K$_{SN}$ and delivers it to the SN over the *Xn-C* interface. The MN forms the S-K$_{SN}$ by associating a Secondary Cell Group (SCG) counter along with the AS security context. The MN sends the SCG counter value to the UE via Radio Resource Control (RRC), signaling the UE needs to generate a new S-K$_{SN}$.

8.9.3 5G Identifiers

There are new identifiers in 3GPP 5G network. These are SUPI and SUCI.

SUPI refers to a Subscription Permanent Identifier. It is a primary identifier in 5G and forms the foundation for all the key derivations together with the subscribers' unique K key. The serving network authenticates the SUPI during the authentication and key agreement (AKA) procedure between UE and network. Furthermore, the serving network authorizes the UE through the subscription profile obtained from the home network based on the authenticated SUPI.

Table 8.3 The 5G keys.

Key	Description
K, CK, IK	Authentication keys.
CK', IK'	Authentication keys derived from CK and IK if EAP-AKA' applied.
K_{AUSF}	Key for AUSF in home network. AUSF generates K_{AUSF} from the authentication material (K, CK and IK for 5G AKA, or from CK' and IK' in case EAP-AKA' is used).
K_{SEAF}	Anchor key. AUSF and ME derive it from K_{AUSF}.
K_{AMF}	Key for AMF. K_{AMF} is derived from K_{SEAF}.
K_{NASint}	Key for NAS signaling. AMF and ME derive it from K_{AMF}. Protects NAS signaling by applying integrity algorithm.
K_{NASenc}	Key for NAS signaling. AMF and ME derive it from K_{AMF}. Protects NAS signaling by applying encryption algorithm.
K_{N3IWF}	The N3IWF receives KN3IWF from the AMF, and uses it as the key MSK for IKEv2 between the UE and N3IWF in the procedures for untrusted non-3GPP access.
K_{gNB}	Key for gNB. AMF and ME derive it from K_{AMF}. gNB and ME derive it further for performing horizontal or vertical key derivation.
K_{RRCint}	Key for RRC signaling. gNB and ME derive it from K_{gNB}. Protects RRC signaling with integrity algorithm.
K_{RRCenc}	Key for RRC signaling. gNB and ME derive it from K_{gNB}. Protects RRC signaling with encryption algorithm.
K_{UPint}	Key for UP traffic. gNB and ME derive it from K_{gNB}. Protects UP traffic between ME and gNB with encryption algorithm.
K_{UPerr}	Key for UP traffic. gNB and ME derive it from K_{gNB}. Protects UP traffic between ME and gNB with integrity algorithm.
NH	Intermediate key. Refers to Next Hop. AMF and ME derive it for providing forward security.

The concealed variant of SUPI is referred to as SUCI. As defined in 3GPP TS 33.501, SUCI is a one-time use subscription identifier that contains the concealed subscription identifier, which may be, e.g. Mobile Subscription Identification Number (MSIN). The SUCI is an optional mechanism managed from the UICC. Its aim is to provide further security in hiding the permanent user identification information. It is a privacy-preserving identifier, and it contains the concealed SUPI.

SUCI is a one-time use subscription identifier that contains the concealed subscription identifier such as the MSIN portion of the SUPI, and additional nonconcealed information needed for home network routing and protection scheme usage. The principle of SUCI is that the UE generates it using a protection scheme with the raw public key that has already been securely provisioned beforehand in control of the home network. Based on the indication of USIM, dictated by the MNO, the calculation of SUCI can be done either by the USIM or the ME. The UE then builds a scheme-input from the part containing a subscription identifier of the SUPI and executes the protection scheme. The UE would not conceal the home network identifier though, such as Mobile Country Code (MCC) or Mobile Network Code (MNC). Please note that there is no requirement for protecting the SUPI in case of unauthenticated emergency call.

In 3GPP-defined 5G system, subscription credentials refer to a set of values stored in the USIM and the ARPF, consisting of at least one or more long-term keys, and the subscription identifier SUPI. These credentials are used to uniquely identify a subscription and to mutually authenticate the UE and 5G core network [17].

8.9.4 PKI Scheme

The public key infrastructure (PKI) scheme is used as a basis for the 5G identifier protection. As stated in ETSI TS 133.501, V. 15.1.0, Chapter 5.2.5 (Subscriber Privacy), the home network public key is provisioned by the home network and it is stored in USIM, while the private key is stored in ARPF. The provisioning as well as updating of the public key is controlled by the home MNO. Likewise, the subscriber privacy is controlled by the home MNO.

The SUPI will not be transferred over the radio interface in clear text. The only exception of this rule is the routing information such as MCC and MNC, which need to be visible.

In case the USIM does not contain yet provisioned public key, the SUPI protection in initial registration is not possible. Null-scheme is applied in this case instead, and it is performed by the ME.

8.9.4.1 UE

In 5G, the UE consists, as has been in previous generations, the ME and secure element, which may be the traditional UICC or evolved variant of it, including embedded and integrated UICCs. Together, the UICC and ME forms the UE.

The security aspects of the UE are described in the 3GPP TS 33.501.

8.9.4.2 UICC and USIM

The ideas of the UICC and USIM in 5G are still the same as has been the case since 3G systems. The original version of the SIM was introduced in Global System for Mobile Communications (GSM), and along with the UMTS, the 3GPP 3G networks developed the concept further by providing the 3G-specific SIM functionality in a form of USIM application that resides in the physical element, UICC based on the smart card standards of the ISO/IEC 7816 and its evolved variants.

As has been the case in 3G and 4G systems of 3GPP, the USIM shall reside on a UICC in 5G. As in the previous release, the UICC may be removable (the "traditional" form factor, e.g. 2FF, 3FF, or 4FF), or nonremovable (i.e. embedded element, or eUICC).

In addition to the continuum of the original UICC in 5G era, 3GPP aligns the new UICC definitions with ETSI. More concretely at the moment, ETSI SCP is working on Smart Secure Platform (SSP), which is a new secure element for the 3GPP system. The SSP can be included as another solution where the USIM can reside on, if the SSP is defined in the Release 15 time frame and if it complies with the security requirements defined in the present document. It should be noted that for non-3GPP access networks, USIM also applies in those cases of terminal with 3GPP access capabilities.

Furthermore, if the terminal supports 3GPP access capabilities, the credentials used with EAP-AKA′ and 5G AKA for non-3GPP access networks shall reside on the UICC [17]. It also should be noted that at present, EAP-AKA′ and 5G AKA are the only authentication methods that are supported in UE and serving network, as dictated by 3GPP TS 33.501.

Table 8.4 The network functions of 5G taking part to the security procedures.

NF	Description
SEAF	The SEAF can initiate an authentication with the UE during any procedure. This takes place upon signaling establishment with the UE. The UE uses SUCI or 5G-GUTI in the registration request procedure. When the SEAF so wishes, it can invoke authentication procedure with the AUSF. Either SUCI (TS 33.501) or SUPI (TS 23.501) is used during authentication. The primary authentication and key agreement procedures bind the anchor key K_{SEAF} to the serving network. The SEAF receives the K_{SEAF} from the AUSF upon a successful primary authentication procedure in each serving network. The SEAF then generates K_{AMF} and hands it to the AMF. The SEAF is co-located with the AMF.
AMF	After successful authentication, key setting takes place as soon as the AMF is aware of the subscriber's identity (5G-GUTI or SUPI). 5G AKA or EAP AKA' results in a new K_{AMF} that is stored in the UE and the AMF. The AMF receives K_{AMF} from the SEAF or from another AMF. The AMF then generates K_{gNB} to be transferred to the gNB; NH to be transferred to the gNB, and possibly NH key to be transferred to another AMF. The AMF also generates K_{N3IWF} to be transferred to the N3IWF.
AUSF	The AUSF refers to authentication server function. During the authentication run, an intermediate key K_{AUSF} is generated which can be left at the AUSF, which optimizes the signaling if the UE registers to different serving 3GPP or non-3GPP networks. This procedure is comparable to fast reauthentication in EAP-AKA', but the subsequent authentication provides weaker guarantees than an authentication directly involving the ARPF and the USIM.
UDM	When UDM receives an authentication request and SUCI, the UDM/SIDF is invoked. The SIDF de-conceals SUCI to obtain the SUPI for the UDM to process the authentication request. The UDM/ARPF selects the authentication method based on the SUPI and subscription data.
ARPF	The ARPF refers to Authentication Credential Repository and Processing Function. It stores, along with the USIM, the 5G subscription credentials, which are the subscription identifier SUPI and long-term key(s). They serve as a base for the security procedures take for identifying subscription and for the mutual authentication of the UE and 5G core network.
SIDF	The SIDF manages the de-concealing of the SUPI based on the SUCI. For this, the SIDF relies on the PKI concept's private key, which is securely stored in the home operator's network. The actual deconcealment takes place at the UDM. The SIDF is protected via access rights, so that only home network's element may present the request to SIDF.

8.9.4.3 Network Functions of Secure Architecture

The network functions taking part on the security procedures with the UE and USIM are SEAF, AMF, AUSF, UDM, ARPF, and SIDF. Table 8.4 summarizes these functions in terms of the security procedures as interpreted from the 3GPP TS 33.501.

8.9.5 Security Procedures

The 3GPP 5G specifications define renewed security procedures. One of the key technical specifications for these is the TS 33.501 (Security Architecture and Procedures for

5G System in Release 15), and TS 33.401 (3GPP System Architecture Evolution; Security Architecture in Release 15). The following subsections discuss these documents considering the latest advances in the SA3, CT6, and other relevant 3GPP work groups dealing with security of the 3GPP network, UICC and USIM.

The authentication procedure consists of the subscribers' unique master authentication key K that is stored in the USIM and in the network. This key together with the ciphering key (CK) and integrity key (IK) (or CK' and IK') are the base for the primary AKA procedures. This procedure enables mutual authentication between the UE and the network. Its tasks also include the production of so-called keying material, i.e. keys that are used for the further security procedures between the UE and network. The keying material provides an anchor key, K_{SEAF}, which is provided by the AUSF of the home network to the SEAF of the serving network.

K_{SEAF} serves for the generation of keys for further security context so that there is no need for a new authentication run, which can happen either in 3GPP network as well as in non-3GPP network. This optimizes the security signaling.

This initial procedure also results in an intermediate key K_{AUSF}. One option for further handling of this key is to leave it into the AUSF, depending on operators' policies.

8.9.5.1 gNB

The ETSI 133 501 V.15.1.0 details the security requirements of the 5G base station (gNB), accompanied by the additional relevant 3GPP technical specifications. Some of the key requirements for the security aspects of the gNB include:

- gNB supports user data and RRC ciphering with UE.
- SMF controls the security policies, and gNB activates the user data ciphering accordingly.
- gNB supports mandatory ciphering algorithms, which are: NEA0, 128-NEA1 and 128-NEA2, while the support of 128-NEA3 is optional.
- gNB may use optional confidentiality protection for the user and control data, the standards encouraging their use as appropriate by the regulation.
- gNB supports integrity protection and replay protection of user and control data with UE, based on the policies dictated by the SMF.
- gNB supports mandatory integrity protection algorithms, which are: NIA0, 128-NIA1, and 128-NIA2, while the support of 128-NIA3 is optional.
- The actual utilization of the integrity protection for user data is optional, though, and if it is used, NIA0 is not allowed. The reason for the latter is that the NIA0 adds medium access control (MAC) overhead between gNB and UE yet without additional security benefits.
- Signaling needs to be integrity protected by default (with some exceptions indicated in 3GPP TS 38.331).

8.9.5.2 Initial Authentication

Figure 8.13 shows the high-level signaling in the initial authentication procedure of 5G, as detailed in 3GPP TS 33.501.

In the security procedures of the 3GPP 5G signaling, the SUPI is not transferred in clear text over 5G radio access network, excluding the delivery of routing information such as MCC or MNC.

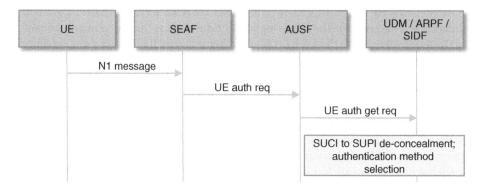

Figure 8.13 The initial authentication procedure of 5G.

The home network public key is stored in the USIM. It should be noted that if the home network has not provisioned the public key in USIM, the SUPI protection in initial registration procedure is not provided, and instead, the ME uses null-scheme. When the UE constructs the SUCI, it contains data fields from which one is the home network public key identifier representing a public key provisioned by the Home Public Land Mobile Network (HPLMN). In case of null-scheme being used, this data field is set to null.

It is also worth noting that for unauthenticated emergency calls, privacy protection for SUPI is not required.

8.9.5.3 5G AKA Authentication

Figure 8.14 shows the 5G AKA authentication signaling flow. It is an enhancement for the enhanced packet system (EPS) AKA procedure as it provides a proof of successful UE authentication from visited network.

There are two types of AV defined in 5G AKA, called 5G HE AV (Home Environment AV) and 5G AV. The 5G HE AV is received by the AUSF from the UDM/ARPF as an authentication info response, and it consists of RAND, AUTN, XRES*, and KAUSF. The 5G AV, in turn, is received by the SEAF from the AUSF, and it consists of the RAND, AUTN, HXRES*, and KSEAF.

8.9.5.4 EAP-AKA′ Authentication

Figure 8.15 summarizes the security signaling for the EAP-AKA′ option. This authentication method is specified in the IEEE RFC 5448 and adapted to the 3GPP specification.

For UE authentication to 5G network via an untrusted non-3GPP access network, the 3GPP TS 33.501 shows the respective signaling. This method uses a vendor-specific EAP method referred to as EAP-5G, utilizing expanded EAP type and the existing 3GPP vendor-ID, registered with IANA under the Structure of Management Information (SMI) Private Enterprise Code registry. The EAP-5G method is used between the UE and the N3IWF and is utilized for encapsulating non-access stratum (NAS) messages. The method is executed between the UE and AUSF. For more details, please refer to the 3GPP TS 33.501, section "Security for non-3GPP access to the 5G core network."

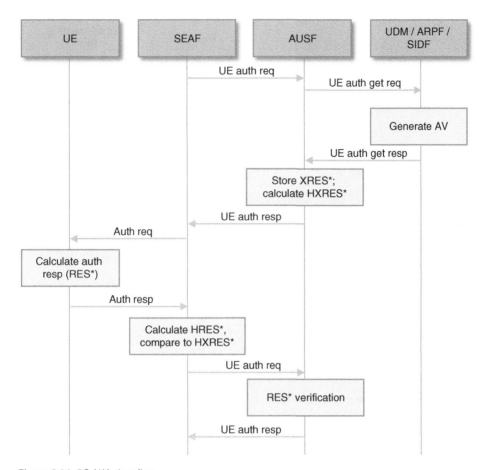

Figure 8.14 5G AKA signaling.

8.10 UICC Evolution

8.10.1 General

The mobile communications networks have been evolving rather slowly since the first days when the SIM card was presented in the GSM system. Ever since, SIM card has had the same idea and removable form until the first decade of 2000. Then, SIM card started to evolve gradually, first introducing Java concept for the compatible applications, or applets. The form factor evolved gradually in such a way that in addition to the original, credit card-sized (which was rarely used) and stamp-sized 2FF, also smaller form factors 3FF and 4FF were introduced into the markets. Also, embedded variants started to appear into the markets driven by small devices such as wearables. The ETSI standardized versions are the MFF1 and MFF2, from which the latter is the only one in use in practice.

For the removable variants, the UICC is a secure base for subscription and its management. It is based on ISO/IEC 7816 smart card standard and it was adopted into the GSM and later, into 3GPP and some other mobile communication systems.

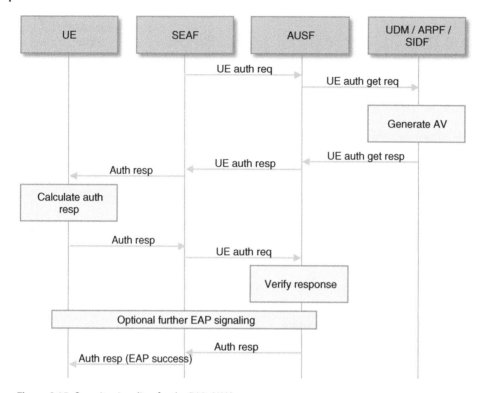

Figure 8.15 Security signaling for the EAP-AKA'.

There are thus both removable and soldered elements in the markets for consumer and machine-to-machine (M2M) communications. For the M2M environment, both embedded and removable variants are used. Some of the typical environments for such M2M modules are the automotive (which requires long time for the element in harsh conditions) and Internet-connected devices, such as intelligent sensors and other IoT devices.

The form of the UICC evolves further as we approach the 5G era, so other variants might also appear in the markets such as further integrated iUICC, which is a fully embedded subscription's functionality within the SoC (system on chip) (Figures 8.16–8.18).

As for the terminology of the subscription and its management needed for the studying of the forthcoming sections, Table 8.5 summarizes some of the key solutions.

8.10.2 SW-Based Security

Due to expected, increasing security threats extending into the wireless environment, the 5G security requires a facelift compared to the previous mobile communications generations. Thus, the typical statements indicate that the security of the 5G needs to be at least, if not better, than that of its predecessors. In addition to the native 5G network security measures, there may be room for also additional layer of security.

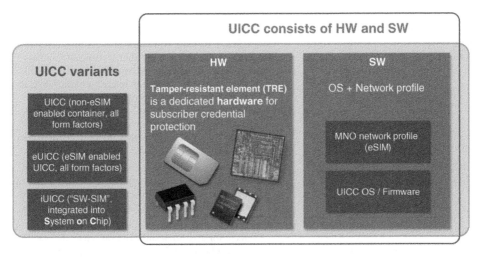

Figure 8.16 The main UICC variants.

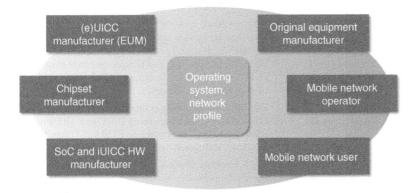

Figure 8.17 The high-level ecosystem for the UICC variants.

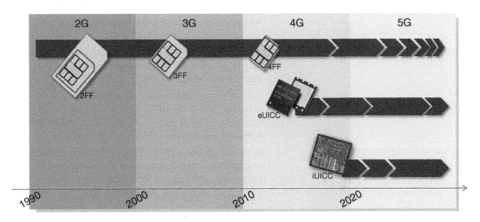

Figure 8.18 UICC variants and respective timelines.

Table 8.5 Key terminology related to the mobile subscription.

SIM Subscriber identity module	Original, removable SIM card as of 2G for storing MNOs and MVNOs subscriber credentials to access the services. Even it is GSM-specific, the term is still widely used in practice when referring to the UICC in mobile communications systems. It is based on the ISO-7816 identification card standard. 2FF, 3FF, and 4FF are the form factors currently in use, and it is typically delivered either within the credit card –sized ID-1 frame or pre-inserted into the mobile device.
UICC Universal integrated circuit card	Removable, tamper-resistant HW-module as defined by ETSI and 3GPP for storing the subscriber data as of 3G systems; It contains subscription data for different cellular systems (like GSM-SIM for 2G and USIM for 3G).
eSIM Embedded subscriber identity module	Concept for activating subscription "virtually," e.g. by scanning QR code (physical or eCard) and triggering Over-the-Air setup procedure. May rely physically on any form factor of UICC or eUICC. GSMA defines eSIM as RSP-capable device.
eUICC Embedded universal integrated circuit card	eSIM-enabled, soldered UICC. MFF2 is standardized, and there are vendor-specific FFs. eSIM and eUICC are often understood as synonyms for soldered SE. GSMA defines eUICC as a tamper-resistant chip, embedded in a device, hosting a FW obeying to the GSMA RSP requirements
iUICC Integrated UICC	The UICC functionality is integrated within the SoC HW. GSMA defines integrated UICC as a TRE within a SoC hosting a FW obeying to ETSI UICC standards or GSMA RSP specifications.
SoC System on chip	Single integrated circuit that incorporates analogue and digital subsystems. It may provide the entire electronics (excluding external nonsilicon components and peripheral hardware) of a complete system, so it could be a single IC system for a mobile handset or tablet computer. The SoC may also include memory for processing firmware and software applications.
SiP System in Package	A single package of electronics for a complete system, typically provided with ICs and components mounted on a single miniature circuit board and encased in epoxy compound, shielding or other packaging.
RSP Remote SIM provisioning	A GSMA specified method to remotely manage the subscriptions of the UICC, whether it is removable or embedded. RSP concept is being considered also as an option to manage the iUICC's MNO profiles.
TRE Tamper-Resistant Element	A silicon enclave within a tamper-resistant chip (TRC) or a SoC that supports the security and tamper resistance requirements for a primary platform.
OFL Open Firmware Loader	Previously named PBL (Primary Boot Loader), OFL acts as a base for the OS downloading for the iUICC's SoC.

Nevertheless, the most economical IoT devices may suffice with the reduced security level, if the protected assets are not particularly valuable.

8.11 5G Security Development

The 5G security aspects and solutions are discussed and planned in multiple forums, including NGMN, 3GPP SA3, and SIMalliance 5G WG. In particular, the progress of the 3GPP LTE-Advanced is paving the way for the most updated and novel 5G security aspects. It is thus expected that several use cases intended for 5G can also be partly covered by LTE.

As for one of the most prominent use cases, the mIoT requires new security solutions. It should be noted that the NB-IoT is a subset and migration path to mIoT, as described by 5G.

Also, a very important driver for the creation of new 5G network security and authentication solutions are the pilots of MNOs, OEMs, and chipset vendors. The new 5G functionalities and their impact on the security need to be evaluated carefully, including the network slicing, network function virtualization, and embedded or integrated UICCs.

8.12 UICC Variants

8.12.1 UICC

The UICC is still a valid security platform in 5G. In fact, the 5G standards designed by ETSI and 3GPP do allow more variety of secure element types compared to the previous generations. The SIM card initiated the practical era for the subscription credentials in form of removable element that has served in the 3GPP services, and later also in other market segments up to now.

The idea of the UICC would not change in 5G, thus, although there will be updates. Thus, it may be that the previous variants of the UICCs won't be compatible with 5G devices. It can be assumed that the removable 5G-UICC will be compatible with the devices pertaining to previous generations yet limited to only those functionalities the systems support based on the earlier specifications.

The file structure of the 5G UICC, thus remains the same as it has been along the route from the very first commercial GSM network launched in Finland back in 1991. ETSI has been the responsible party maintaining the up-to-date specifications related to the UICC in 3GPP networks, and this joint activity continues. As the terminology of the standardized SIM elements has been renewed in 2017, the UICC and is variants are now referred to as SSP, ensuring the wider set of forms of the traditional idea of the original SIM and UICC cards. Specifically, for the integrated UICC, or iUICC, the ETSI-specific term is nowadays integrated Smart Secure Platform (iSSP).

8.12.2 eUICC

The embedded variant of the UICC, i.e. eUICC, refers to an element that is permanently soldered into the device – whether it is consumer device or M2M equipment.

The main standardization bodies related to the eUICC are:

- GSMA, which has designed various versions of specifications for the subscription management of the eUICCs
- GlobalPlatform
- SIMalliance

Some examples of the eUICC subscription management specifications defined by GSMA are the V2 for the consumer environment, and V3 for the M2M environment. The aim of the GSMA is to converge these specifications to cover all the use cases within a single set of specifications referred to as V4.

It should be noted that the previously utilized term of the GSMA for the subscription management, RSP (remote SIM provisioning) is nowadays replaced by the term eSIM, which refers to the generic electrical form of the SIM.

8.12.3 vUICC

The virtual UICC is a nonstandard term and it may refer to the so called "soft SIM" in general, or other similar variants such as eSIM. In this book, the term vUICC is not utilized by default in order not to cause fragmented interpretations on the topic.

8.12.4 iUICC

The latest addition into the variants of the UICC family is the integrated variant of the UICC, iUICC.

8.12.4.1 3GPP Security Requirements for iUICC

The UICC-related requirements have been largely unchanged as for the HW since the early days of the SIM card. The UICC has been thus an isolated, well-protected, and tamper-resistant element, which stores the data and keys. Along with the new potential solutions, the 3GPP has adjusted the requirements leaving room for further innovation. One of the respective requirements states that the subscription credentials need to be protected using a tamper-resistant secure HW in such a way that the actual data may be stored *also* outside of tamper-resistant component.

This new statement means that the SoC HW can perform encryption of the subscription-related data, such as operating systems and MNO profile, while the data can be stored, e.g. in an external or separate nonvolatile memory (NVM). This makes it feasible for the iUICC to act as a security chip to handle the subscription data.

The iUICC concept is being standardized for 5G. The 3GPP is targeting to have the first stage of the 5G defined by the Release 15 providing its technical specifications by March 2018. Thus, in 5G, a dedicated UICC is not mandated any more, but it can be still used as a secure element (SE).

8.12.5 Benefits and Role of iUICC

The ecosystem has been set up globally for the UICC concept, including certified security environment processes for the UICC providers, so it continues being a highly feasible solution for the consumer markets. Nevertheless, the iUICC would provide connectivity that may be especially feasible for low-cost and small IoT devices. The iUICC is thus aimed to provide an additional form for the current UICC portfolio.

The iUICC reduces the required physical space as the SoC integrates the UICC functionality into it, and there is benefit via reduced energy consumption and reduced number of components because the processing capability and memory elements of the SoC can be used for housing the subscription data, too.

8.12.6 iUICC Standardization

The removable UICC remains the foundation for 5G security. It is a proven, tamper-resistant, hardware-based secure element, which protects user credentials. The UICC provides a significant level of protection being compliant with, e.g. the GSMA Security Accreditation Scheme (SAS).

The eUICC refers to a secure element permanently soldered into a device such as mobile phone or an IoT/M2M device, such as a smartwatch or connected thermostat. Managing and updating eUICC user data, up to and including a change to the MNO providing connectivity for a device, must be accomplished without the need to remove and replace a traditional SIM card. The GSMA, GlobalPlatform, and SIMalliance organizations have been standardizing methods for this remote UICC management, and the key specifications are now commonly referred to as eSIM (known also as remote SIM provisioning or RSP). The clarification of these definitions and capabilities will be a driving factor to future successes with 5G connectivity.

The iUICC is the newest addition to the portfolio. Integration embeds the UICC functionality further into the SoC, such as the base band or application processor of the device. Due to this deeper integration, there is no need for a separate physical secure element. Optimization of both space utilization and energy consumption may prove beneficial but will require a new ecosystem for managing its contents.

The standardization of the next generation UICC is managed by ETSI under the SSP, which covers UICC development independent from form factors; the current classes are embedded Smart Secure Platform (eSSP), which refers to the embedded SSP (eUICC), and iSSP, which refers to the integrated SSP (iUICC). There may be more classes introduced in the future. As for the evolution of the UICC toward the eSSP and iSSP concepts, Table 8.6 summarizes the standardization bodies and the key roles for establishing different variants of the expanding UICC family.

For subscription management (remotely managing the contents of the embedded and integrated elements), standardization ensures the market will not fragment due to vendor-specific solutions. A consistent solution ensures interoperability in downloading the MNOs profiles, UICC operating systems and subscriber data into the protected containers of the UICC. The user can thus activate the MNO profile (one at a time), change the profile between MNOs and delete the profile whether located in a home country or roaming.

The subscription management standardization is still under development. The first solutions are available for the M2M and consumer use cases based on the GSMA specifications, while work continues toward a converged solution via a single platform.

Table 8.6 The main stakeholders for standardization of UICC evolution.

Stakeholder	Role on UICC	Work item description
ETSI TC SCP	iSSP, eSSP	ETSI SCP (Smart Card Platform) works on SSP (Smart Secure Platform) concept, successor to the UICC, without specified form factors, considering products such as integrated SSP (iSSP) and embedded SSP (eSSP).
GlobalPlatform	OFL, VPP	The eSSP, and the lower level operating system of the SoC for iSSP can be upgradeable via the Open Firmware Loader (OFL) and Virtual Primary Platform (VPP) of GlobalPlatform. VPP aims at abstracting the hardware and easing the firmware development.
GSMA	M2M, eSIM	The OS downloading of eSIM can be done via GSMA standards, and relying on ETSI SCP's SSP, covering the management of MNO specific OS and profile bundles.
SIMalliance	UICC, eUICC, eSE	Defines the ecosystem for OFL covering the initial loading of OS, and refurbishment (reload of OS) at OEM factories, for eUICC, UICC, and eSE.
3GPP	USIM, 5G	3GPP CT6 Group will work on integrating 5G features in the specifications. CT6 works on 5G USIM based on SA3 solutions.

8.12.7 The UICC in the 5G Era

The "traditional" UICC will maintain its relevance as a tamper-resistant, hardware-based, secure element to securely store keys and other confidential information. There is an extensive legacy device base and established ecosystem, which has been formalized for the manufacturing, logistics, and personalization of the MNO profiles. The UICC provides an adequate security level compliant with strict international accreditation requirements as dictated by the GSMA SAS and payment institutions.

With rapid developments in the IoT and the need for built-in security for IoT devices, the market will benefit from a deeper integration of the USIM functionality into these devices. An ultra-small device would not be able to house a traditional UICC form factor due to its physical size. Thus, increasingly tiny form factors of the UICCs are expected to arrive on the 5G markets, including a variety of both embedded UICCs and integrated UICCs.

As depicted in the beginning of this chapter, in Figure 8.1, additional UICC variants are expected in the 5G era along with the previous form factors – which are still worthwhile in the established ecosystems and provisioning techniques. For the embedded and integrated UICC variants, standardization defines their management via the interoperable eSIM concept.

The new interoperable variants of the UICC are facilitated by ETSI, which has widened the traditional definition of a UICC. At the same time, the terminology now reflects an improved generic form, and, as a result, ETSI refers to the evolution of UICC by the new term SSP. It should be noted that SSP is not equivalent to UICC, and UICC still continues to exist. Nevertheless, certain evolutions will end up in UICC next version and others will be included in SSP such as iSSP. We can thus expect to see more variants of secure elements soon.

For both the embedded form of the UICC and the integrated variant, referred to as eSSP and iSSP, respectively, per new ETSI terminology, the management of the respective MNO profiles and subscription data can be based on the GSMA eSIM concept and supported by the ecosystem developed by GlobalPlatform and SIMalliance.

As has been the principle in 3GPP releases, backward compatibility is essential and will continue to be with the new variants so the impacts of 5G on the original UICCs are expected to be minimal.

There will be 5G-specific files included in the new UICCs as of Release 15. Nevertheless, the 5G file structure remains the same as in previous versions of the USIM. In addition, the pre-existing UICC form factors are still valid, referring to the standardized 2FF, 3FF, 4FF, as well as to the MFF2 used in M2M applications.

8.12.8   UICC Profiles in 5G

3GPP defines the 5G subscription credentials as a set of values in the USIM and the ARPF. 5G subscription credentials consist of long-term key or set of keys K, unique for each user, and the SUPI, which uniquely identifies a subscription. The K and SUPI function for mutually authenticate the UE with the 5G core network.

The subscription credentials are processed and stored in the USIM. In addition, 3GPP has confirmed the USIM still resides on a UICC in the 5G era.

The 3GPP Technical Specification TS 31.102 outlines the characteristics of the USIM application. In Release 15, there is one new file that defines the configuration parameter for handover between WLAN and EPS of 4G. This file is a new service in the universal SIM Toolkit (UST). There is also a modification to the specification for enhanced coding of Access Technology in EFPLMNwAcT (user-controlled mobile network selector with access technology file) to accommodate 5G Systems (5GSs).

The 3GPP technical specification TS 31.103 outlines the features of the IP Multimedia Services Identity Module (ISIM) application. In Release 15, there is one new file defined in this specification with a configuration parameter for handover between WLAN and EPS. There is also an updated file ID for EF-XCAPConfigData (IP Multimedia Subsystem [IMS] configuration file) in the ISIM, to match the value used in the USIM. Please note that the MCPTT (mission-critical push-to-talk) corrections already considered in Release 14 are updated in Release 15. In addition, the Estimated_P-CSCF_Recover_Time timer is removed, which requires an update on data in EFIMSConfigData file.

8.12.9   Changes in Authentication

In general, the same principles apply as previously defined for both the file structure and actual files throughout prior 3GPP generations.

There will be some impacts on the previously defined UICC, but these impacts are not overwhelming while a majority of the new functionality will be handled by the ME. Changes will also bring new profile files and a few modifications with the aim to maintain UICC compatibility. As an example, if there is a network roaming priority list file OPLM-NwACT present, and it contains 5G network content, which may not be available yet, there are no issues as the device passes through the remaining options for prioritizing the available accessed networks.

As has been the case previously, the USIM shall reside on a UICC also in 5G. The UICC may be removable or nonremovable. The ETSI SCP is working on a new secure element referred to as SSP. The SSP can be included conditionally as another solution on which the USIM can reside, but only if the SSP is defined in the Release 15 time frame, and if it complies with the security requirements defined in the present specifications.

For non-3GPP access networks, USIM applies in the case of a terminal with 3GPP access capabilities. If the terminal supports 3GPP access capabilities, the credentials used with EAP-AKA and 5G AKA for non-3GPP access networks shall reside on the UICC.

As a summary, 3GPP TS 33.501 currently specifies only the requirements for secure storage and processing of subscription credentials. However, per the interim agreement in 3GPP Technical Recommendation TR 33.899, the two accepted solutions are UICC and SSP.

For further information on the subject, the reference document 3GPP TS 33.501 also presents examples of how additional authentication methods can be used with the EAP framework.

8.12.10   5G UICC Security

The principles designed for managing the UICCs, as defined via the eSIM construct (formerly known as RSP of GSMA), will also apply in 5G. The evolution of 5G subscription management includes the convergence to support consumer and M2M devices via a common platform which is still under development. A good source of information for the advanced subscription management valid for 5G can be found in Ref. [18].

The evolution of the UICC toward embedded and integrated variants needs to be considered when designing the proper security as the iUICC security solution necessarily changes established workflows. As International Telecommunications Union (ITU) IMT-2020 dictates, the level of the 5G security shall be at least the same or better than in previous telecommunications systems.

The integrated UICC provides deeper physical integration into the SoC, making security breaches, such as local side channel attacks, even more difficult to execute. Concurrently however, there may be unknown, potential security weaknesses opened due to the new technology, especially regarding the intercommunication between the shielded and secured area and external electronics, such as NVM. Hence, there is a need to protect the solution via up-to-date accreditation schemes.

8.12.11   Legacy UICC Compatibility with 5G

The 3GPP TR 31.890, V15.1.0, Section 8.2 states that it is expected that the UE might be able to access the 5G system with the existing UICCs that are found already in commercial market. It is thus possible that the UE will camp on a 5G system even if the UICC is not able to recognize the new network. As the3GPP specifications aim to provide a good level of backward compatibility in general, it is important to ensure the functionality of this scenario by updating the specifications where appropriate and feasible.

In practice, to handle this scenario, a new bit is required in the USIM Service Table to indicate the support for the 5G system so that the ME can adjust its behavior depending on the UICC support. Furthermore, a new bit is needed in the Terminal Profile to provide the ME with possibility to inform the UICC of the 5G support.

This abovementioned requirement results in the need to assess the impact on the fields defined for the USAT (USIM Application Toolkit) in the Release 15 of 3GPP TS 31.111 [19], to ensure the mapping of the information is correct between the UICC and ME. These fields are:

- Access Technology
- Location Information
- Timing Advance
- Network Measurement Results
- Network Rejection Event
- Bearer Independent Protocol
- Call Control on Packet Data Unit (PDU) Connection Establishment by USIM
- Data Connection Status Change Event
- Launch Browser

The USAT provides mechanisms that allow applications, which reside in the UICC, to interact and operate with any ME supporting the specific mechanisms that are required by the application. These mechanisms include, e.g. profile download, data download to UICC, user's menu selection, USIM call control, USIM's mobile-originated short message control, event download, etc. The current USAT specification includes various scenarios and technical possibilities such as proactive UICC commands between the ME and the UICC, and the specification forms an important base for the utilization of the USIM.

The abovementioned requirements to accommodate 5G for the use scenario of the previous USIMs was taken as a work item of 3GPP System Architecture (SA3) work group, and it is documented in Ref. [20]. In the earlier version of the 3GPP TS 31.111 (clause 8.49), the definition of Bearer object refers to General Packet Radio Service (GPRS), UMTS Terrestrial Radio Access Network (UTRAN) packet service and Evolved UTRAN (E-UTRAN). This specification for the bearer service was expanded to refer to Next Generation Radio Access Network (NG-RAN) as well, in practice by using the existing table to include an encoding of "03" as per the agreement in TR 31.890 clause 8.2.9.

8.12.12 iUICC Components

Figure 8.19 presents one of the potential solutions for the iUICC and its subscription management. It should be noted that this solution may or may not be reality, as its details are under discussion in standardization bodies.

Figure 8.20 show more detailed picture on the respective technical solution.

The iUICC HW refers to a tamper-resistant enclave on a SoC, based on separate security controller. This concept is designed to replicate the functions of the UICC in such a way that no separate component – "traditional" SIM forms or embedded SIMs – are needed due to the integrated functions in the SoC itself.

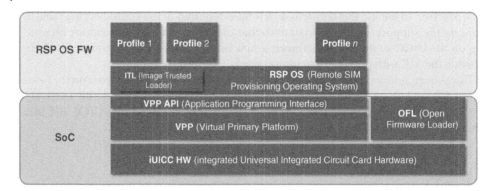

Figure 8.19 Potential solution for the iUICC subscription management.

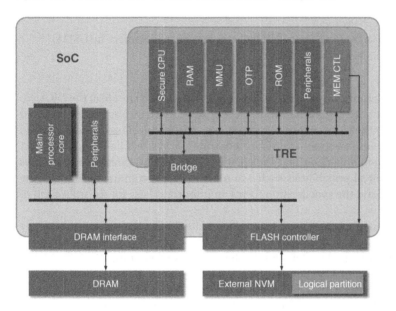

Figure 8.20 A potential iUICC solution based on the tamper resistant area of the SoC.

As the SoC in question, such as base band processor, does not by default have enough memory for any additional data required for the UICC functionality (such as operators subscription profile and UICC's operating system), the iUICC can be implemented by relying on a NVM so that the data is stored there in encrypted form. Based on the service organization control (SOC), it is possible to deploy default safety mechanisms, such as Tamper-Resistant Element (TRE) counter, against rollback attacks (Figures 8.21 and 8.22).

8.12.13 Components for iUICC Data Loading

The Virtual Primary Platform (VPP) shields the HW-specifics under a standardized API and enables OS-interoperability on SoC platforms. The VPP ensures that the embedded UICC manufacturer can still develop OS for different customers in a standardized

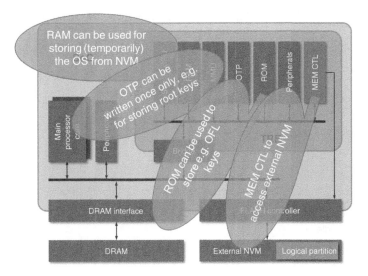

Figure 8.21 Part of the existing architectural components of the system on chip (SoC) can be reutilized for storing securely the operating system and profile of the mobile subscription.

Figure 8.22 The potential position of the VPP and other functional elements of iUICC concept. Please note the iUICC concept is still under standardization, and e.g. ETSI has included it as a work item under the term iSSP.

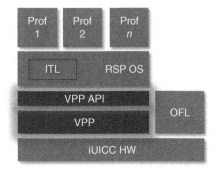

environment. The VPP also ensures that once developed, the OS runs everywhere. This, in turn, provides means for shorter time to market and reduced testing efforts.

Along with the VPP, MNOs can maintain relation with trusted OS suppliers, obtaining specific features. On the other hand, the VPP API enables the OS manufacturer to develop and provide an OS without need to access any HW-specific or proprietary information.

The Open Firmware Loader (OFL) provides means for interoperable OS loading. In addition to the VPP itself, an open loading mechanism for trusted OS providers is mandatory. It enables OS providers to load their firmware (which consists of the OS and data) on any HW platform (including the iUICC) in an interoperable and standardized way. The OFL of the GlobalPlatform is thus the initial step to standardize such loading mechanism.

The GSMA RSP provides smooth migration of iUICC. It protects the investment in eSIM management infrastructure and processes. RSP is possible in an interoperable and standardized way. It paves the way for an evolution toward OS/image management for

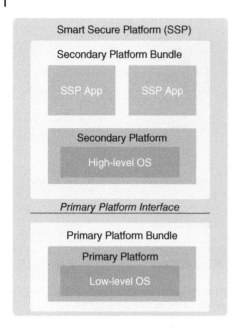

Figure 8.23 The SSP concept as defined by ETSI TS 103 465 V.1.1.0.

iUICC. Furthermore, the RSP OS provides interoperable evolution from eSIM to iUICC management.

ITL (Image Trusted Loader) loads the OS either fully or partially keeping the existing data. The ITL is proprietary technology of each OS manufacturer.

At present, ETSI is working on SSP bundle, aligned with 3GPP Release 16. Figure 8.23 summarizes the high-level idea of the iSSP architecture, based on ETSI TS 103 465 V.1.1.0.

References

1 GSMA, 2017, "iUICC Assurance report," GSMA, 2017.
2 3GPP, 2016, TR 22.891 Feasibility Study on New Services and Market Technology Enablers, Stage 1, 2016-06.
3 3GPP, 2016, Study on Architecture for Next Generation System, August 2016.
4 N. Alliance, 2016, 5G security recommendations Package #2: Network Slicing, April 2016.
5 N. Alliance, 2016, Description of Network Slicing Concept, January 2016.
6 E. G. 0. NFV, 2013, Network Functions Virtualisation (NFV); Architectural Framework, October 2013.
7 "openbts.org/," 2016. [Online]. Available: http://openbts.org. [Accessed September 5, 2016].
8 N. Alliance, 2016, 5G security - Package 3: Mobile Edge Computing/Low Latency/Consistent User Experience, July 2016.
9 3GPP, 2016, Study on Enhanced General Packet Radio Service (EGPRS) access security enhancements with relation to cellular Internet of Things (IoT), June 2016.

10 I. S. G. M. E. Computing, 2014, Mobile-Edge Computing – Introductory Technical White Paper, September 2014.

11 E. G. M. 003, 2016, Mobile Edge Computing (MEC); Framework and Reference Architecture, March 2016.

12 3GPP, 2016, TR 22.864 Feasibility Study on New Services and Markets Technology Enablers – Network Operation, June 2016.

13 SIMalliance, Technical Whitepaper, 2016, 5G Security – Making the Right Choice to Match your Needs, October 2016.

14 3GPP, 2018, TS 33.220 Generic Bootstrapping Architecture, September 2018.

15 GSMA, 2016, Solutions to enhance IoT Authentication using SIM Cards (UICC), October 2016.

16 C. Dietze, 2016, "x"UICC: Evolution of the UICC is turning into a Revolution, April 2016.

17 3GPP, 2018, "TS 33.501," 3GPP, 2018.

18 GSMA, 2018, "eSim Whitepaper: The what and how of Remote SIM Provisioning", March 2018 https://www.gsma.com/esim/wp-content/uploads/2018/06/eSIM-Whitepaper-v4.11.pdf. Accessed 13 November 2018.

19 3GPP, 2018, "Universal Subscriber Identity Module (USIM); Application Toolkit (USAT)," 3GPP, April 2018.

20 3GPP, 2018, "3GPP 31.11 Change Request CR 0674 (C6-180066); Enhance Bearer object in Toolkit spec to accommodate 5GS (15.1.0)," 3GPP, 2018.

9

5G Network Planning and Optimization

9.1 Overview

This chapter discusses 5G network planning and optimization by giving an overview to the planning methods and processes applicable from the 4G era, new considerations for 5G, the key requirements for 5G planning, including flexibility and high reliability, need for multiple network and radio access technology layers, high mobility and connectivity between networks and devices, precise positioning, and high security.

This chapter provides generic guidelines for the 5G core network dimensioning, including capacity estimation and reservation, dynamic capacity allocation, and cloud-aided methods. This chapter also details aspects on 5G radio network planning, including coverage, capacity, and parameter planning, evolved radio link budget (RLB) and its adjustment based on use cases.

9.2 5G Core and Transmission Network Dimensioning

As has been the case in all the previous mobile networks, the essential dimensioning tasks remain the same in 5G, too. The main elements in the optimal network planning are related to the capacity, coverage, and quality of service (QoS). The balancing of these parameters depends on the wanted costs; in the optimal network, the return of investment is a balance, not too expensive as it wastes the capacity, not too low cost as the customers' churn increases along with the reduced quality.

9.2.1 Capacity Estimation

Along with the expected huge amount of Internet of Things (IoT) devices that will surge within forthcoming years communicating simultaneously in always-on mode, the 5G networks will need to considerably support increased traffic capacity. Not only will capacity demand increase but also the traffic patterns will diversify to include evolved Multimedia Broadband (eMBB), critical communications, and massive IoT traffic mixed within the same network.

Nevertheless, the 5G system has been designed in such a way that the network slices take care of each traffic type individually, and the hardware resources are only utilized as needed for the network functions (NFs). Furthermore, the network slices per use case

5G Explained: Security and Deployment of Advanced Mobile Communications, First Edition.
Jyrki T. J. Penttinen.

are designed so that they only include the needed network functions, which further optimizes capacity utilization.

The 5G networks are remarkably different from the previous generations due to the virtualization and service-based architecture model. In the 5G core network dimensioning, some of the practical questions are thus:

- How to plan the strategy for the slices, i.e. what network functions are needed in each use case
- How to reserve enough hardware processing power
- How the transmission capacity should be dimensioned for peak and average traffic
- How to prepare for the unexpected traffic peaks

The 5G networks are designed for a completely new era with considerably increased data rates, number of simultaneously connected IoT devices, low latency, and a variety of use cases that are possible to optimize via the network slicing concept differentiating characteristics of the connection [1].

The 5G tackles the near-future need by providing evolved means for network scalability and flexibility. The key technologies for this are network functions virtualization (NFV) and software-defined networking (SDN).

The planning of 5G thus requires renewed dimensioning models for cost-optimized deployments, handling a variety of different traffic demand scenarios. Not only is the radio interface dimensioning of utmost importance, but also the optimal placement of the core network elements is essential for providing adequate performance for critical communications, with the lowest possible latency values. Thus, the location of physical edge computing elements, i.e. data centers (DC), must be considered.

NFV offers the needed flexibility, as it removes the HW dependency yet enables the means for fast deployment and service updates. NFV can thus contribute positively to the cost optimization of the network deployments and service offering.

SDN, in turn, decouples data and control planes (CPs) of network functions. It provides open application programming interfaces (API), which is in practice based on OpenFlow protocol, for the decoupled planes.

The modeling and optimization for the NFV and SDN are related to the physical location of SDN controllers and switches, Virtual Network Function (VNF) resource allocation, and placement (Figure 9.1).

As described in [2], considering the resource allocation and placement of VNFs, a mathematical model can be formed to find an optimal placement for virtual core gateways handling sporadic traffic increase, e.g. when a large crowd event takes place. It also is possible to adapt machine-learning techniques to find an optimal placement for VNFs as a function of data center resources. Some examples of such research can be found in [3–5].

In the initial phase of 5G network deployment, the most typical scenario is to rely on the already existing legacy core network of 4G. In the 4G Evolved Packet Core (4G EPC), the user and control plane functions are based on dedicated hardware and respective software for each function. Examples of these elements are the Proxy Gateway (P-GW), Serving Gateway (S-GW), Home Subscription Server (HSS), and Mobility Management Entity (MME). In the 5G core network, instead, functions that solely handle the control plane, such as the MME, may be deployed based on VNFs. This refers to the concept that allows the use of a common HW, and its resources are shared dynamically for all

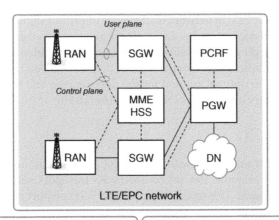

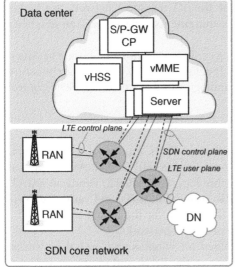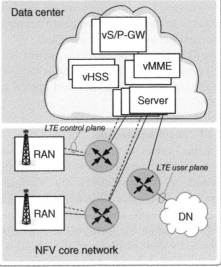

Figure 9.1 Evolution of the core network for the LTE/EPC, SDN, and NFV.

the virtualized network functions on a cloud infrastructure. For the functions taking care of both user and control planes, such as S-GW and P-GW, the respective traffic management can rely on SDN-based or NFV-based solution.

The SDN-based architecture makes the control plane mobile core functions run as VNFs, whereas the gateway functions of S-GW and P-GW are decoupled into SDN controllers and SDN switches. The SDN controllers are deployed at the data center. They configure the SDN switches handling the user plane (UP) traffic, and they take care of the EPC control plane signaling procedures.

In the NFV-based architecture, the control plane core network and gateway functions, which are S-GW and P-GW, are executed as VNFs at data centers' common HW. In this scenario, the control and user plane processing of the gateways run on common cloud servers. This means that the previously utilized HW of the core network can be replaced by adapting transport switches forwarding control and user plane traffic within radio access network (RAN), data centers, and external networks.

Ref. [2] states that the cost of network load may be improved significantly by distributing the data center infrastructure because as there are more available data centers, also more VNFs may be deployed complying the requirements for latency, to decrease additional SDN control traffic and respective cost of the network load. It should be noted, though, that there is an optimal amount for the data centers, and the values exceeding four do not significantly improve the load cost optimization.

9.2.2 Transport

The dimensioning of transport network infrastructure is one of the important tasks of a 5G operator. As an example, Ref. [6] considers 5G transport network dimensioning for a dense urban scenario. This research shows how 5G transport network architecture based on the 5G-XHaul project can be deployed.

As reasoned in [6], deployment models for the 5G transport network are highly needed to understand the respective capacity requirements and to assess the suitability of current networking technologies.

5G-XHaul is a European-wide project defining converged fronthaul (FH) and backhaul (BH) networks for future 5G mobile networks. It defines a logical transport architecture integrating multiple wireless and optical technologies under a common SDN control plane [7–9].

9.2.2.1 Xhaul Architecture

5G-XHaul refers to a unified transport for flexible RAN functional split concept. The respective consortium is formed by partners such as IHP GmbH, ADVA Optical Networking, Airrays GmbH, Blu Wireless Technology, COSMOTE, Fundació i2CAT, Huawei Technologies, Telefónica I+D, TES Electronic Solutions, and various universities. Ref. [10] presents the European Union (EU)'s plan for the XHaul solution to be used as a base for the transport infrastructure of the 5G networks.

The SDN control plane of the Xhaul is unified for wireless and optical networks. It is aware of demand spatiotemporal variations in the RAN, and interfaces for joint RAN–Transport leg.

The data plane unifies the transport of wireless networks, which in this example are P2MP mm-Wave (60 GHz), Sub-6 GHz networks, optical, Time-Shared Optical Network (TSON), and Wavelength Division Multiplexing Passive Optical Network (WDM-PON).

9.2.3 Effect of Network Slicing

Network slicing is one of the key features in fully functional and optimized 5G. It provides the means to design and deploy customized communication systems and to integrate services from different verticals. It can be expected that there will be a massive amount of simultaneously communicating devices, which require sporadic connectivity and small-scale data transfer. The respective required QoS may vary considerably, depending on the use case. The mIoT services would thus benefit from a highly dynamic, end-to-end slice representing radio access and core network with enhanced control and user planes. Other use cases might require a higher grade of security, whereas still others might want the highest grade of reliability, i.e. lowest possible latency.

The requirements for the slices serving different use cases is thus one of the essential dimensioning tasks of the mobile network operator (MNO), to plan the services reflecting practical requirements of each use case. Some of the optimization tasks may be related to the reduced CP signaling while maximizing UP resource utilization [11].

9.3 5G Radio Network Planning

9.3.1 Radio Channel Modeling

For MNO to plan the radio coverage and respective capacity for the 5G network, the radio channel modeling continues being one of the most essential tasks. There is a variety of models developed and under research, and often times there may be set of different models instead of only a uniform one due to the differences in geographical topologies in global level, and even within a single country. As one example of this variety of strategies, the model might be based on a highly accurate 3D map data of the environment with ray-tracing (RT) principles in the densest city areas, whereas the accurate coverage area prediction is not so essential in the most rural and remote areas. It is a matter of balancing the return of investment, as the expense of providing the most accurate map data is justified in some applications compared to the basic raster data, which is sufficient for the remote locations.

It can be generalized that the radio propagation models have been rather established for the practical needs of the earlier mobile generations up to 4G. As the 5G includes considerably higher frequency bands and may utilize much wider bandwidths and novelty modulation schemes, there have been several research projects investigating the suitable models for the new era. Some examples of the research include the studies carried by the 3GPP and radio section of International Telecommunication Union (ITU-R), as representatives of the standardization and regulation. There also are models provided by METIS, mmMagic, MiWEBA, COSTIC1004, IRACON, and the 5G mmWave Channel Model Alliance. One of the recent references investigating these models is found in [12].

The number of the 5G radio frequency (RF) bands is expected to be much higher than in previous mobile generations, including multiple mm-Wave bands, so the previous models need to be revised. The massive multiple-in, multiple-out (MIMO) antenna technology and hybrid beamforming are important methods in 5G era to achieve highest data speeds aimed to a very large number of users within dense areas. The respective array's gain tackles the radio transmission loss of the higher-frequency bands [13].

For commercial products, not all the antenna elements are supported by separate RF chains. In practice, the arrays may be connected to a baseband unit by only a small number of RF chains, while antenna elements may be divided into subarrays [14]. Thus, each subarray may compose several predefined beam shapes. It is expected that the main directions of the beams will cover an angular sector of interest.

The 3GPP TR 25.996, V14.0.0 contains relevant information as a basis for the MIMO radio dimensioning. It is a useful resource for investigating more thoroughly the modeling and impacts of the key parameters such as radio path arrival angles to the received power level for suburban macro cell, urban macro cell, and urban micro cell topologies.

9.3.2   5G Radio Link Budget Considerations

The base of the RLB is the same as in previous systems; the path loss is estimated based on the key parameters, considering the set of gains and losses.

The base for the coverage area estimate is thus the offered capacity in terms of bit rate and the respective achievable coverage area. The higher the data rate, based on the adaptive code rate of the Orthogonal Frequency Division Multiplexing (OFDM) principle, the smaller the coverage area is, which means that the highest data rates are achieved close to the base station while the lowest bit rates are available still in the edge area of the cell coverage.

The additional aspects for the 5G are a result of novelty techniques such as higher-order MIMO antennas and respective gain per single path. This is a highly dynamic environment, as the number of simultaneous users, including a massive set of IoT devices, causes the available capacity to fluctuate accordingly.

The goal of the initial phase of 5G radio network planning is to obtain a rough estimate for the coverage and capacity within planning area. This can be done by using a radio link budget. It serves as a simple tool for estimating the achievable path loss values per planned environment, between the transmitter and receiver in uplink (UL) and downlink (DL). One of the benefits of this nominal planning phase is to have a high-level estimate of the number of base stations needed to serve the planned area. The more detailed estimate is obtained posteriorly by utilizing a radio network planning tool and applying more accurate models and detailed digital maps. Both nominal and detailed radio network planning are essential phases, as the radio network may have a major impact on the expenses of the deployment, and optimization of the base station locations, utilized power levels, antenna heights, and other practical aspects, the return of investment can vary considerably.

In a practical network planning, the predesigned base station locations are selected as candidates within a certain preferred search areas. This plan may not always be realistic, as the "site hunting" may be sometimes highly challenging due to restrictions. As an example, the construction of towers or even installation of pure antenna elements on the walls may be forbidden in certain environments such as historical city centers. Thus, the nominal and detailed plans before the actual deployment may change.

The radio link budget thus gives a rough estimate of the expected usable cell radius, which is practical especially for the nominal radio network coverage planning phase. For the detailed planning, radio network planning tools are typically utilized in such a way that the propagation models and other assumptions are adjusted per cluster type based on the practical field tests.

Figure 9.2 shows the principle of the 5G radio link budget for the estimation of the maximum usable path loss between the 5G NodeB (gNB) and 5G user equipment (UE), which indicates the cell radius both in uplink and downlink. By applying the path loss value into adequate radio propagation model, the cell radius can be estimated for different geographical types such as dense urban, urban, suburban, rural, and open area.

The balancing of the uplink and downlink is considered important for the two-way use cases such as Internet Protocol (IP) voice call that requires similar performance in both directions when aiming to achieve a fluent user experience in the cell edge region (i.e. in all the circumstances, both A and B subscriber hear each other). For the balancing of less critical cases, the data rates in uplink and downlink depend on the application. Thus, for

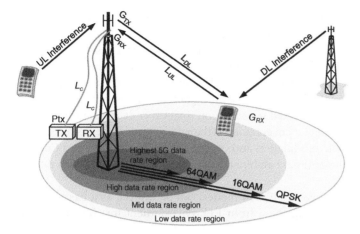

Figure 9.2 Principle of the radio link budget parameters.

downloading of data such as web pages, the data speed in uplink is not the limiting factor, so even considerably lower data rates in uplink still provides the adequate interpretation of the fluent performance.

The downlink and uplink of 5G use different modulation schemes that each provide different level of performance, which, in turn, converts into variable cell radius values. In addition to the received useful carrier signal, the final radio link performance depends on the level of interfering signals.

The key parameters in the downlink direction are the output power level of the gNB transmitter (P_{TX}), the cable and connector losses (L_c), and the transmitting antenna gain (G_{TX}). The role of the adaptive antenna systems (AASs) and beamforming technologies is more important in 5G compared to previous generations. The optimization of the radio links per user is thus more detailed in 5G. Furthermore, the cable loss can be assumed to be compensated by the antenna system's front-end as the gNB's transmitter is typically located into the antenna module, and the respective transmission form the gNB up to the antenna system is delivered via practically lossless fiber optics.

The radiating power (P_{EIR}, effective isotropic radiating power) refers to the radiating power level for the omni-radiating antenna. In practice, the mobile communication system is often relying on directional antennas, or in case of the 5G, on further optimized, beam forming with highly directional link so the antenna gain G_{TX} is taken into account accordingly to represent the realistic power level.

In the receiving end of the downlink, the key parameters of the 5G UE are the sensitivity (S), receiver antenna gain (G_{RX}) and noise figure (NF). In typical cases, the 5G terminal's antenna gain can be assumed to be either 0, or even negative, depending on the level of integration of the components and device's antenna size.

In the uplink, the key parameters are the transmitting power of the 5G UE, the antenna gain of the terminal and gNB (which are likely the same as in downlink if the same antenna types are utilized), the cable and connector losses, and the sensitivity of the gNB receiver.

The maximum allowed path loss values L_{DL} and L_{UL} are calculated for different modulation schemes by assuming the minimum functional received power level per coding scheme, and by subtracting the emitted and received powers. The estimate can be done per geographical area type of interest in outdoors, and the indoor coverage can be estimated via typical penetration loss values for buildings.

As a rule of thumb, the most robust modulation schemes, Binary Phase Shift Keying (BPSK) and Quadrature Phase Shift Keying (QPSK), provide the largest coverage areas, but at the same time result in the lowest capacity. The higher constellation Quadrature Amplitude Modulation (QAM) schemes provide the highest data rates, but with reduced coverage areas. Also, the least protected transmission (highest code rates) provides the highest data rates but within smaller coverage areas than the heavier protected codes do.

The 5G system has dynamic modulation and coding (MCS, Modulation and Coding Scheme), which provides the optimal combination of the modulation and code rate at all the times.

The well-known radio path loss prediction models can be utilized as a basis for the 5G coverage estimations, adjusting them from LTE environment for the support of extended frequency bands of the 5G.

The link budget calculation is typically based on a minimum throughput requirement at the cell edge. This approach provides the cell range calculation in a straightforward way. The order of defining the throughput requirement prior to the link budget calculation, it is possible to estimate the bandwidth and power allocation values for a single user, which mimics the behavior of the realistic scheduling sufficiently well for the purposes of link budget calculation.

The 5G link budget can be estimated by analyzing the effect of the key parameters as summarized in Tables 9.1 and 9.2. In these examples, a relatively narrow band of 360 kHz is assumed for the uplink, and equally not aggressive 10 MHz band for the downlink transmission.

The link budget can be planned in the following way, e.g. in the downlink direction with the 10 MHz assumption, as shown in the previous example. The radiating isotropic power (d) can be calculated by taking into account the transmitters output power (a), the antenna feeder and connector loss (b), and transmitter antenna gain (c), so the formula is: $d = a - b + c$.

The minimum received power level (p) of the UE can be calculated as $p = k + l + m - n + o$, utilizing the terminology of the link budgets just shown. The noise figure of the UE depends on the quality of the model's HW components. The minimum signal-to-noise (or, combined signal-to-noise and interference, SINR) value j is a result of the simulations. The sensitivity k of the receiver depends on the thermal noise, g, the noise figure of the terminal, h, and SINR in such a way that $k = g + h + j$.

The interference marginal l of the link budget represents the average estimate of the noncoherent interference originated from the neighboring base station elements. The control channel proportion m also slightly degrades the link budget. In the link budget calculations, the effect of the antenna of the UE can be estimated as 0 dB if no body loss is present near the terminal. For the external antenna, the antenna gain increases respectively, but the logical estimate of the average terminal type is a stick model with only a build-in antenna.

The effect of the data speed can be estimated in a rough level by applying a rule of thumb that assumes a path loss of 160 dB in uplink, using 64 kb s^{-1} data rate. Whenever the bit rate grows, the maximum allowed path loss drops, respectively. A simple and practical assumption is that the doubling of the data rate increases the path loss

Table 9.1 The principle of the downlink radio link budget.

Downlink	Unit	Value
Transmitter, eNodeB		
Transmitter power	W	40.0
Transmitter power (a)	dBm	46.0
Cable and connector loss (b)	dB	2.0
Antenna gain (c)	dBi	11.0
Radiating power (EIRP) (d)	dBm	55.0
Receiver, terminal		
Temperature (e)	K	290.0
Bandwidth (f)	MHz	10.0
Thermal noise	dBW	−134.0
Thermal noise (g)	dBm	−104.0
Noise figure (h)	dB	7.0
Receiver noise floor (i)	dBm	−97.0
SINR (j)	dB	−10.0
Receiver sensibility (k)	dBm	−107.0
Interference margin (l)	dB	3.0
Control channel share (m)	dB	1.0
Antenna gain (n)	dBi	0.0
Body loss (o)	dB	0.0
Minimum received power (p)	dBm	−103.0
Maximum allowed path loss, downlink		**158.0**
Indoor loss		**15.0**
Maximum path loss for indoors, downlink		**143.0**

by 3 dB. This can be assumed to work sufficiently well for selected channel coding and modulation schemes.

As stated in 3GPP TS 38.300, informative Annex B, to improve UL coverage for high frequency scenarios, Supplementary Uplink (SUL) can be configured. Details of SUL can be found in 3GPP TS 38.101, but as a generic principle, with SUL, the UE is configured with two ULs for one DL of the same cell, as depicted on Figure 9.3.

9.3.3 5G Radio Link Budgeted in Bands above 6 GHz

9.3.3.1 Default Values for Presented Scenarios

This section is based on the information derived from the International Telecommunication Union (ITU) feasibility study of International Mobile Telecommunications (3G) (IMT) in bands over 6 GHz [15]. The frequency bands above 6 GHz for small cells are introduced to 5G as a novelty solution. These bands are included in the band set to facilitate the scalability, capacity, and density that 5G requires in the efforts to integrate

Table 9.2 The principle of the uplink radio link budget.

Uplink	Unit	Value
Transmitter, terminal		
Transmitter power	W	0.3
Transmitter power (a)	dBm	24.0
Cable and connector loss (b)	dB	0.0
Antenna gain (c)	dBi	0.0
Radiating power (EIRP) (d)	dBm	24.0
Receiver, eNodeB		
Temperature (e)	K	290.0
Bandwidth (f)	MHz	0.36
Thermal noise	dBW	−148.4
Thermal noise (g)	dBm	−118.4
Noise figure (h)	dB	2.0
Receiver noise floor (i)	dBm	−116.4
SINR (j)	dB	−7.0
Receiver sensibility (k)	dBm	−123.4
Interference margin (l)	dB	2.0
Antenna gain (m)	dBi	11.0
Mast head amplifier (n)	dB	2.0
Cable loss (o)	dB	3.0
Minimum received power (p)	dBm	−131.4
Maximum allowed path loss, uplink		**155.4**
Smaller of the path losses		**155.4**
Indoor loss		**15.0**
Maximum path loss in indoors, uplink		**140.4**
Smaller of the path losses in indoors		**140.4**

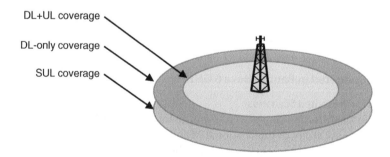

Figure 9.3 The principle of SUL (supplementary uplink).

the expected considerable number of small cells into the rest of the cellular network infrastructure. Based on the propagation characteristics of these high frequencies, with a variety of bands as indicated in Section 5.3.3 of this book, they offer the highly needed additional capacity for the network densification. The drawbacks of these high-frequency bands are the very small coverage area per cell due to their higher path loss values, as well as additional system complexity for the RF front end and for advanced antenna design. Nevertheless, the current technological advances have provided feasible products around 60 GHz, so it can be assumed that there will be respective commercial deployments as 5G matures.

The elemental task of the radio link budget in general is to estimate the received signal power after the propagation loss in radio interface. This can be based in theory on the free space loss equations, as presented in Rec. ITU-R P.525. As the typical mobile communications networks contain variety of environments, including obstacles such as buildings and vegetation, non-line of sight (NLOS) models are essential also in 5G.

ITU-R Report M.2376-0 (07/2015) [15] presents scenarios for frequency bands above 6 GHz. As an example, the reference introduces key system parameters:

- Carrier frequency: 28 GHz
- Transmitter's Effective Isotropic Radiated Power (EIRP) and receiver's antenna gain: 65 dBm
- Bandwidth: 1 GHz
- Receiver's noise figure: 7 dB
- Miscellaneous losses: 10 dB
- Target data rate: 1 or 0.1 Gb/s
- Target signal-to-noise ratio (SNR): 0 or −11.4 dB

These values work as a base for our examples presented in the next sections, too, which summarize the analysis of Ref. [15].

9.3.3.2 28 GHz

Based on the examples presented in [15], the estimate of the path loss in 28 GHz can be done by the path loss models summarized in Table 9.3.

Ref. [15] assumes that the center frequency for this scenario is 28 GHz, with 1 GHz bandwidth. The total sum of transmitter output power, transmitter, and receiver antenna gains for EIRP scenario is assumed to be 65 dB; this result is feasible in practice, e.g. with the values of 30 dBm, +25 dBi, and +10 dBi, respectively. These assumptions, combined

Table 9.3 Selected models for the 28 GHz scenario.

Model	Path loss equation	Notes
Open space	$L_a = 61.4 + 20 \cdot log_{10}(d)$	LOS
Campus	$L_b = 47.2 + 29.8 \cdot log_{10}(d)$	NLOS
Dense urban, option 1	$L_c = 96.9 + 15.1 \cdot log_{10}(d)$	NLOS; for $d < 100$ m
	$L_c = 127.0 + 87.0 \cdot log_{10}(d/100)$	NLOS; for $d > 100$ m
Dense urban, option 2	$L_d = 61.4 + 34.1 \cdot log_{10}(d)$	NLOS

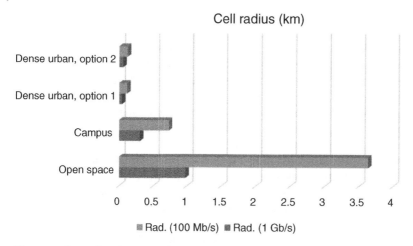

Figure 9.4 Examples of 28 GHz link budget's cell radius as interpreted from Ref. [15].

with the propagation models presented in Table 9.3, result in the maximum cell radius per area type and data rate as summarized in Figure 9.4.

In this example, the low-powered gNB can provide indoor coverage of some tens of meters in dense urban areas (such as New York City Center) up to some hundreds of meters in more open NLOS environments such as semi-open campus locations. The line of sight (LOS) scenario offers 1 Gb/s up to near 1 km, and 100 Mb/s up to about 3.5 km. It should be noted that these values can vary greatly, depending on the more specific parameters, geographical topology, and antenna heights, among other variables, but this example of Ref. [15] gives a rough understanding on the achievable 5G radio coverage on 28 GHz.

9.3.3.3 39 GHz

Continuing with the summarizing of examples presented in Ref. [15], the 39 GHz band is suitable principally in the small-cell deployments in most dense environments as the respective coverage area is impacted. Furthermore, for the adequate accuracy of the model, ray-tracing principles are beneficial.

Table 9.4 presents the formulas used in Ref. [15]. The free-loss formula for L_e is based on the LOS scenario of ITU-R P.2001. The L_f is formed based on the detailed description of Ref. [15] in its Appendix A4.3 (Figure 9.5).

In this scenario, the user equipment's transmit power is 10 dBm, antenna gain 15 dBi, bandwidth 500 MHz, input noise power −80.9 dBm, radio receiver's noise figure 10 dB,

Table 9.4 Selected models for the 39 GHz scenario.

Model	Path loss equation	Notes
Dense urban, based on ITU-R P.2001	$L_e = 92.44 + 20log_{10}(39) + 20log_{10}\left(\frac{d}{1000}\right)$	LOS
Dense urban, street canyon	$L_f = 78.28 + 23.6log_{10}\left(\frac{d}{5}\right)$	NLOS

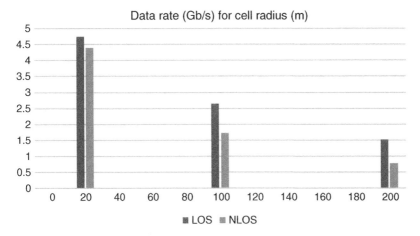

Figure 9.5 The achievable data rates per selected distances on 39 GHz band.

and implementation loss 10 dB. The transmitter output power is 19 dBm, transmit antenna gain +24 dBi provided by beam steering.

It should be noted that the model presented in Ref. [15] is simplified, and it does not consider advanced baseband techniques like coding gain effects on the channel model. Nevertheless, this information is valuable to understand the expected data rate as a function of cell range in 39 GHz band.

9.3.3.4 60 GHz

Continuing with the scenarios presented in Ref. [15], Table 9.5 summarizes the path loss equations for the 60 GHz frequency band. As in the previous section, the free-loss formula for this scenario, L_g, is adjusted based on LOS scenario of ITU-R P.2001, and the L_h is formed based on the detailed description of Ref. [15] in its Appendix A4.3.

In this scenario, the user equipment's transmit power is 10 dBm, antenna gain 15 dBi. Unlike in the previous scenario for 39 GHz, bandwidth of 2 GHz is now applied. The input noise power remains at −80.9 dBm, as well as radio receiver's noise figure 10 dB, and implementation loss 10 dB. The transmitter output power is 19 dBm. Unlike in the previous case, a higher-gain transmit antenna of +29 dBi is applied as there is such commercial potential for highly directive point-to-point installations.

Even the frequency range as such causes more attenuated path loss at 60 GHz compared to 39 GHz. Applying the parameters provided in Section 9.3.3.1, the coverage area can be greatly enhanced.

Table 9.5 Selected models for the 60 GHz scenario.

Model	Path loss equation	Notes
Dense urban, based on ITU-R P.2001	$L_g = 92.44 + 20log_{10}(60) + 20log_{10}\left(\frac{d}{1000}\right)$	LOS
Dense urban, street canyon	$L_h = 82.02 + 23.6log_{10}\left(\frac{d}{5}\right)$	NLOS

Table 9.6 Selected models for the 72 GHz scenario.

Model	Path loss equation	Notes
Dense urban, based on ITU-R P.2001	$L_i = 69.59 + 36.7log_{10}(d)$	NLOS
Dense urban, street canyon	$L_j = 69.59 + 28.6log_{10}(d)$	LOS

9.3.3.5   72 GHz

Continuing with the examples, Table 9.6 presents the models for 72 GHz link budget. In this example, the antenna array may be more dense and more compact. This higher antenna gain of +34 dBi may be achievable at this frequency range.

In this scenario, the transmitter power is assumed to be +26 dBm, transmitter EIRP +60 dBm, and receiver antenna gain +21 dBm. The resulting coverage at 1 Gb/s is indicated to be around 120 m when NLOS is applied. For LOS conditions, the coverage at 1 Gb/s may be in range of 470 m.

References

1 Głąbowski, M., Gacanin, H., Moscholios, I., and Zwierzykowski, P. "EditorialDesign, Dimensioning, and Optimization of 4G/5G Wireless Communication Networks," 2017.

2 Basta, A., Blenk, A., Hoffmann, K. et al. (2017). Towards a cost optimal design for a 5G mobile core network based on SDN and NFV. *IEEE Transactions on Network and Service Management* 14 (4).

3 Gebert, S., Hock, D., Zinner, T. et al. (2014). Demonstrating the optimal placement of virtualized cellular network functions in case of large crowd events. *ACM SIGCOMM Computer Communication Review* 44 (4): 359–360.

4 Fukushima, M., Hayashi, M., and Chawuthai, R. et al., "LawNFO: A decision framework for optimal locationaware network function outsourcing," Proc. 1st IEEE Conf. Netw. Softwarization (NetSoft), London, U.K., London, 2015.

5 Shi, R., Zhang, J., and Chu, W. et al., "MDP and machine learning-based cost-optimization of dynamic resource allocation for network function virtualization," Proc. IEEE Int. Conf. Services Comput. (SCC), New York, NY, USA, 2015.

6 Demirkol, I., Camps-Mur, D., Bartelt, J., and Zou, J. "5G Transport Network Blueprint and Dimensioning for a Dense Urban Scenario".

7 EU, "5G-XHaul Project, Deliverable 2.2, System Architecture Definition," 5G-Xhaul Project, 1 071 2016. http://www.5g-xhaulproject.eu/download/5G-XHaul_D_22.pdf. [Accessed 30 July 2018].

8 5G PPP, "5G PPP Architecture Working Group, View on 5G Architecture," 01 July 2016. https://5g-ppp.eu/wpcontent/uploads/2014/02/5G-PPP-5G-Architecture-WP-July-2016.pdf. [Accessed 30 July 2018].

9 "Dynamically Reconfigurable Optical-Wireless Backhaul/Fronthaul with Cognitive Control Plane for Small Cells and Cloud-RANs," 5G-Xhaul, 01 07 2016. http://www.5g-xhaulproject.eu/index.html. [Accessed Jan 2017].

10 European Commission, 5G-XHaul, 2.2, System Architecture Definition, Horizon 2020, European Union funding for Research and Innovation, 1 July 2016. https://www.5g-xhaul-project.eu/download/5G-XHaul_D2_2.pdf. [Accessed 30 July 2018].

11 Trivisonno, R., Condoluci, M., An, X., and Mahmoodi, T. "mIoT Slice for 5G Systems," Design and Performance Evaluation.

12 Kyösti, P. (2018). *Radio Channel Modelling for 5G Telecommunication System Evaluation and Over the Air Testing*. Oulu: University of Oulu.

13 Larsson, E.G., Edfors, O., Tufvesson, F., and Marzetta, T.L. (2014). Massive MIMO for next generation wireless systems. *IEEE Communications Magazine* 52 (2): 186–195.

14 Verizon, "Physical layer procedures, Technical Report TS V5G.213 v1.0," Verizon 5G TF, 2016. http://5gtf.net/V5G_213_v1p0.pdf. [Accessed 30 July 2018].

15 ITU, "Report ITU-R M.2376-0, Technical feasibility of IMT in bands above 6 GHz," ITU, July 2015.

10

Deployment

10.1 Overview

The 5G system (5GS) has raised a big interest in the telecommunications industry, and it can be expected that the markets will be deploying early-stage 5G well before the actual ITU-R IMT-2020 candidate technologies have been presented and formally approved.

These early deployments represent still pre-5G era as the International Telecommunications Union (ITU's) point of view is considered. Nevertheless, Release 15 being the first 3GPP version, and preliminary candidate for the IMT-2020 process, it is a matter of markets to decide how the 5G deployments are called in practice. Based on the strictest interpretations, only Release 16 may fully comply with the ITU requirements and thus the earliest possible timing for such respective and complete 5G networks can be seen as of the year 2020.

10.2 Trials and Early Adopters Prior to 2020

10.2.1 Verizon Wireless

Verizon has been active in 5G network trialing with mobile equipment and network infrastructure providers and has produced a set of 5G system specifications for testing the 5G performance. The focus of these trials has been on a subset of New Radio (NR), such as Fixed Wireless Access without mobility support. During this time, the user devices have been limited to Consumer Premise Equipment, such as set-top boxes. The main aim of this activity has been to provide more efficient last mile wireless connectivity, and to collect experiences on the expected performance of the final 5G.

The Verizon 5GTF has been in elemental role for the 5G trialing. It has been a cooperative setup with Cisco, Ericsson, Intel, LG, Nokia, Qualcomm, and Samsung participating in the efforts. The idea has been to define a common and extendable platform for the 28 and 39 GHz fixed wireless access trials and deployments.

The focus of these efforts has been on creation of 5G radio interface specification between the user equipment (UE) and the network promoting interoperability among

5G Explained: Security and Deployment of Advanced Mobile Communications, First Edition.
Jyrki T. J. Penttinen.
© 2019 John Wiley & Sons Ltd. Published 2019 by John Wiley & Sons Ltd.

network, UE, and chipset manufacturers, thus facilitating early 5G trials. Ref. [1] details more information on the setup of the Verizon 5G Technical Forum and respective specifications.

10.2.2 AT&T

Another example of the early 5G activities in the United States is the AT&T trialing with Ericsson, Samsung, Nokia, and Intel. The focus of this activity has been on fixed wireless 5G trials, including business and residential customers in Waco, Texas; Kalamazoo, Michigan; and South Bend, Indiana, by the end of 2017. The applied use cases included car wash and other small businesses, apartment test sites, etc.

The aim of this activity has been to obtain insights into mm-wave performance and propagation, foliage, building materials, device placement, topology, and weather impacts on the radio link. The results have demonstrated experimented data speeds up to 1 Gb/s and latency rates well under 10 ms for the radio link. The setup has been utilized to conduct further outdoor prestandards mobile 5G testing.

10.2.3 DoCoMo

DoCoMo has been active in testing 5G concept in Tokyo. The utilized bands have been 4.5 and 28 GHz bands, with data rates up to 20 Gb/s per base station.

There has also been a demo basing on current understanding on 3GPP 5G NR, in cooperation with NTT DOCOMO. The achieved coverage was noted to be up to 2.0 km when 39 GHz was utilized. The tests have resulted 1 Gb/s peak throughput for a single user, which has been demonstrated to work for real-time 4K video conference. The list of the DoCoMo trial can be found in [2].

10.2.4 Telia Company

Also, Telia Company has been active on proofing the 5G performance. There has been a joint roadmap with Ericsson and Intel to let Telia customers experience 5G services in Tallinn and Stockholm [3].

10.2.5 European Union

European Union coordinates various academic and industry initiatives via 5GPPP initiative. The respective roadmaps can be found in Ref. [4].

10.3 5G Frequency Bands

As 5G era is approaching, there are global efforts to make suitable frequency bands available for the new era. Regulators are thus actively exploring the national strategies to open up new frequencies. One of the most important occasions in this sense is the ITU-R WRC-19, which discusses the international strategies to dedicate 5G frequencies per region. According to [5], some of the possible bands can be identified according to Table 10.1 (status in December 2016).

Table 10.1 Snapshot of 5G-related initiatives.

Region	Strategy
USA	The FCC (Federal Communications Commission) aims to enable 5G over various bands in the states. Some examples are auction of the 600 MHz band, and a CBRS band in 3.5 GHz with 150 MHz bandwidth to be shared amongst unlicensed generic access and licensed stakeholders. There is also approximately 11 GHz of spectrum (28, 37–40, and 64–71 GHz) for mmWave applications, amongst other candidate bands for IMT-2020.
Europe	The EC (European Commission) has identified RF bands of 700, 3.4–3.8, and 24.25–27.50 GHz for the use of 5G. EC aims to facilitate the wide-scale deployment of 5G across Europe by 2020, initially covering major cities while the full deployment in urban areas and major terrestrial roads is covered by 2025.
China	China has carried out trials in 3.4–3.6 GHz band aiming to extend the bands to cover 3.3–3.4 and 4.8–4.99 GHz. China also has studied the use of mmWave bands such as 24.25–27.5 and 27.5–29.5 GHz.
South Korea	The early 5G would rely on mmWave bands in 28 GHz. The target of this initiative was trial at the 2018 Winter Olympic Games. Other options in South Korea include 37.5–40 GHz band.
Japan	The planned bands include 3.6–4.2, 4.4–4.9, and 27.5–29.5 GHz.
Australia	The focus is in 3.4–3.7 GHz combined with potential mmWave bands.

10.4 Core and Radio Network Deployment Scenarios

The 3GPP TR38.801 presents practical deployment scenarios for the 5G architecture. These scenarios are:

- Noncentralized
- Co-sited with E-UTRA
- Centralized
- Shared radio access network (RAN)
- Heterogenous deployment

As an alternative point of view, the high-level scenarios can be categorized into two cases: non-standalone (NSA) and standalone deployments, which are described in Annex J of 3GPP TR 23.799. The following sections summarize some key scenarios found from the 3GPP specifications.

10.4.1 Noncentralized

The noncentralized deployment scenario refers to a set of 5G NodeB (gNB) elements equipped with a full protocol stack. The scenario is suitable especially in macro cell and indoor hotspot deployments. The gNB elements can connect with other gNB elements or LTE eNB and eLTE eNB elements. The eLTE eNB is the evolved version of eNB that supports connectivity to both 4G enhanced-packet core (EPC) and 5G Next Generation Core (NGC). Figure 10.1 depicts the principle of this scenario.

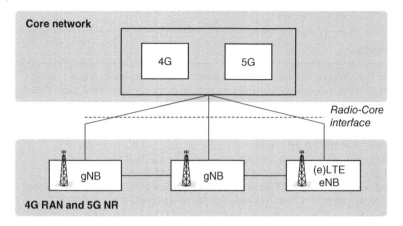

Figure 10.1 Noncentralized deployment scenario.

10.4.2 Co-sited with E-UTRA

In co-sited deployment scenario, the 5G NR functionality is co-sited with 4G E-UTRA functionality as part of the base station or as multiple base stations at the same site.

This deployment scenario is suitable for any cell types, including Urban Macro. Load balancing works to optimize the spectrum resource utilization for the radio access technologies, e.g. in such a way that the lowest frequencies can be applied for ensuring the constant coverage in cell edge, while higher frequencies ensure the optimal user data rates (Figure 10.2).

10.4.3 Centralized Deployment

The 5G NR supports centralization of the upper layers of the NR radio protocols. As depicted in Figure 10.3, various gNB elements can be attached into the centralized unit via transport. The performance of the transport layer between the lower layers of

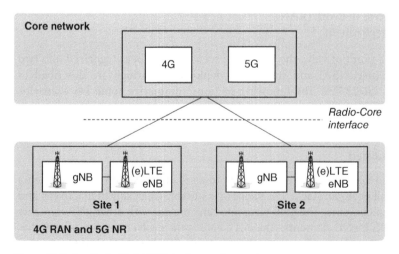

Figure 10.2 Co-sited with E-UTRA deployment scenario.

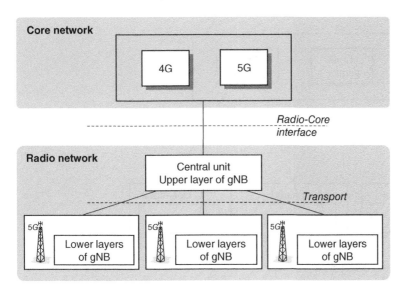

Figure 10.3 Centralized deployment scenario.

gNB elements and centralized element may be set differently depending on the case. Non-co-sited and co-sited deployment scenarios with E-UTRA are applicable in this scenario. These cases include:

- *High-performance transport*, such as optical network, and lower layers of gNB elements. This option can provide benefits for advanced coordinated multipoint (CoMP) scenarios and for optimization of scheduling, which provides high capacity.
- *Low-performance transport*, with less-demanding performance in terms of bandwidth, delay, synchronization and jitter, and lower layers of gNB elements. This scenario is possible as the higher protocol layers have lower performance requirements for the transport layer.

10.4.4 Shared RAN

The 5G NR supports shared RAN deployment scenario applicable for the cooperation between various core network operators. In addition to the shared serving areas, which can vary from localized up to national coverage, the operators can have also their individual serving areas adjacent to the shared ones.

The 5G NR supports shared RAN scenarios where various hosted core operators are present. It is also possible to deploy heterogenous RAN service areas, which can be optimal for locations such as indoor deployments in cooperation. This is due to the requirement of a fluent interworking between shared RAN and nonshared RAN (Figure 10.4).

10.4.5 Heterogeneous Deployment

A heterogeneous deployment refers to a set of more than one scenarios described previously in Sections 10.4.1–10.4.4 within the same geographical area.

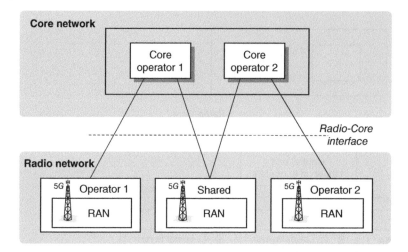

Figure 10.4 Shared RAN deployment scenario.

10.5 Standalone and Non-Standalone Deployment Scenarios

The 5G system provides the mobile network operators (MNOs) with various deployment options to pave the way for the gradual evolvement of 5G. The key aspect in this is the possibility to rely on the 4G infrastructure while the fully native 5G infrastructure is under construction. The native 5G refers to standalone network architecture where the 5G NR is directly connected to the 5G core, i.e. NGC. The non-standalone options refer to different scenarios where 4G and 5G radio access technologies, i.e. E-UTRA and NR, respectively, as well as 4G and 5G core networks cooperate.

As stated in 3GPP TS 23.799, the most relevant scenarios for deploying the interfaces between the radio access and core networks are the following:

- Option 2: The Standalone 5G NR gNB is connected to the 5G NGC.
- Option 3/3A, the LTE eNB is connected to the 4G EPC with non-standalone NR. The NR user plane (UP) is connected to the EPC via the LTE eNB (Option 3) or directly (Option 3A).
- Option 4/4A, the 5G NR gNB is connected to the 5G NGC with non-standalone E-UTRA. The E-UTRA user plane is connected to the NGC via the gNB (Option 4) or directly (Option 4A).
- Option 5, the eLTE eNB is connected to the NGC; Please note that this option has been deprioritized in the latest 3GPP specifications.
- Option 7/7A, the eLTE eNB is connected to the NGC with non-standalone NR. The NR user plane connection to the NGC goes via the eLTE eNB (Option 7) or directly (Option 7A).

These deployment scenarios have been described in detail, e.g. in the 3GPP TR 23.799, Annex J for further information.

The interfaces shown in the following sections refer to:

- *NG1* is designed for the control plane between the 5G UE and 5G Core.
- *NG2* is designed for the control plane between the 5G RAN and 5G Core.
- *NG3* is designed for the user plane between the 5G RAN and 5G Core.

10.5.1 Standalone 4G/5G RAN and NGC

The standalone deployment scenario refers to the 4G E-UTRAN or 5G NR and 5G NGC, which are connected directly with each other while the 4G infrastructure does not participate in the 5G communications.

Figure 10.5 depicts the principle of this scenario as interpreted from the 3GPP TS 23.799. In this scenario, the *NG1* interface is the control plane reference point for 5G UE and 5G core. The *NG2* interface is the control plane reference point for the 5G NR and 5G core while the *NG3* interface is the user plane reference point for the 5G NR and 5G core.

The scenario with 4G E-UTRAN connected to the 5G NGC is similar to the 5G NR and 5G NGC, meaning that it is called standalone because there are no new requirements needed for the support of legacy EPC or E-UTRAN.

Both scenarios are depicted in Figure 10.5. This scenario with newly defined 5G NR radio access and 5G NGC is relevant when NR access systems are deployed in areas without the presence of legacy Long Term Evolution (LTE) and EPC infrastructure.

10.5.2 Non-Standalone 5G NR with 4G EPS

The 4G E-UTRA can be used in a dual connectivity architecture in such a way that the 4G eNB represents an anchor component carrier for user and control plane. The 5G NR access represents thus a secondary component carrier and uses solely control plane. This scenario refers to non-standalone architecture because legacy eNB and EPC network functions are needed for the deployment of the user plane of the NR access.

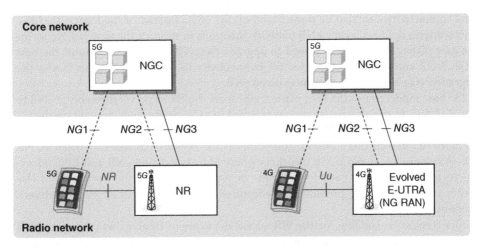

Figure 10.5 Standalone deployment scenario for 5G and 4G, referring to respective NR or evolved E-UTRA connected to the next-generation core, as defined in 3GPP TR 23.799.

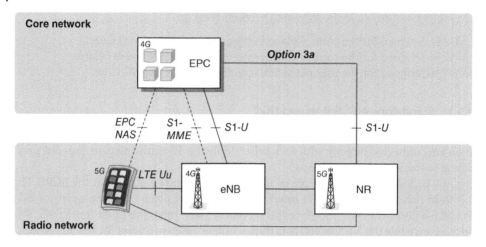

Figure 10.6 Non-standalone 5G NR and 4G EPS as defined in 3GPP TR 23.799. The additional *S1-U* interface as indicated in the figure refers to Option 3a, whereas the scenario without it is referred to as Option 3.

This scenario can be divided into two parts: Option 3 refers to the architecture that delivers the user plane traffic from the NR access via the 4G eNB, which routes it to EPC. In Option 3a, the UP traffic can be routed directly from the 5G NR access node to the 4G EPC by relying on the *S1-U* interface.

This scenario is depicted in Figure 10.6. It is suitable in early deployment of 5G NR when legacy eNB and EPC infrastructure is available, and the deployment can thus be based on the already existing *S1-MME, S1-U,* and *NAS* (non-access stratum) signaling interfaces.

10.5.3   Non-Standalone 5G NR with 5G NGC

The non-standalone scenario with E-UTRA supporting 5G NR user plane, providing dual connectivity, can also be done based on 5G NGC instead of the legacy 4G EPC. This scenario requires legacy eNB network functions to support the new NR access for the user plane. Also, there is a need to upgrade the eNB elements to support the new *NG1, NG2,* and *NG3* interfaces for the connectivity to the 5G NGC. This is why the upgraded 4G eNB is referred to as evolved eNB.

The base model routes the user plane traffic from the NR radio via the evolved eNB to the 5G NGC. In the alternative option referred to as 7a, traffic goes directly via the 5G NR access node to the 5G NGC.

Figure 10.7 depicts this scenario. This scenario may be worth considering when 5G NR access systems are deployed in locations where already existing eNB and EPC infrastructure exists and are ready for upgrade or replacement with the evolved eNB elements and the 4G NGC, providing new features with these new or upgraded network elements.

This scenario is suitable for the NR access deployment in areas where legacy eNB elements and EPC are prepared for the upgrade or replacement, the evolved eNBs taking over together with 5G NGC. These provide new set of features as a basis for more advanced services in paving the way for the native 5G elements.

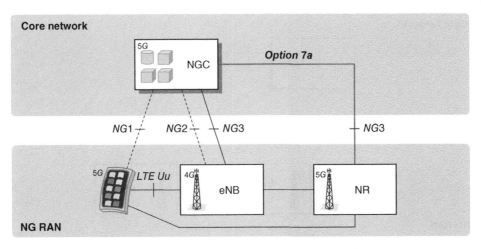

Figure 10.7 Option 7 of non-standalone 5G deployment, referred to as Option 7. The scenario with the additional *NG3* interface is referred to as Option 7a.

10.5.4 Non-Standalone 4G E-UTRA with 5G NGC

Yet another deployment scenario is based on dual connectivity with the 5G NR access node acting as the anchor component carrier for user and control planes. The evolved eNB has a role of secondary component carrier and transfers only user plane data. Signaling to the 5G NGC takes place via the *NG1*, *NG2*, and *NG3* interfaces.

In this base-scenario referred to as Option 4, the user plane traffic goes from the evolved eNB radio access via the 5G NR access and ends at the 5G NGC.

This mode is called non-standalone because the evolved eNB user plane communications require access to the anchor point of the 5G NR. As is the case in previous options, this one has an alternative Option 4a, which means that the user plane traffic goes directly from the evolved eNB to the 5G NGC relying on the *NG3* interface. Figure 10.8 depicts Options 5 and 5a.

This scenario provides fluent deployment of 5G NR elements in locations where already existing 4G eNBs and 4G EPC infrastructure exists and can be upgraded or replaced with Evolved eNB elements and 5G NGC in such a way that the 5G NR access would be the dual connectivity anchor point instead of 4G. One such case may be when the 5G NR can utilize lower-frequency bands than the evolved 4G eNB elements.

10.5.5 Deployment Scenarios of gNB/en-gNB

The 3GPP TS 38.401, Annex A, presents informative deployment scenario for the gNB and en-gNB. Based on this information, Figure 10.9 shows logical nodes (CU-C, CU-U, and DU), internal to a logical gNB/en-gNB. In this figure, the protocol terminations are indicated with circles.

The *NG* interface is defined in 3GPP TS 38.410, while the *Xn* interface is defined in 3GPP TS 38.420 and the *F1* interface is defined in 3GPP TS 38.470.

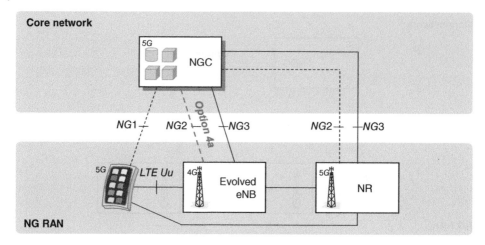

Figure 10.8 The non-standalone scenario representing Option 4 and 4a, the latter being of the same architecture as Option 4, but with the additional *NG2* interface between eNB and NGC. As is the case with other options, this one is defined in 3GPP TR 23.799.

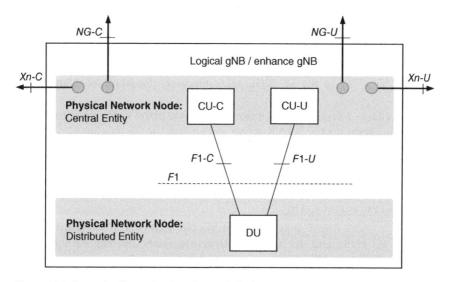

Figure 10.9 Example of logical gNB and en-gNB deployment.

ETSI 133 501 V. 15.1.0, Chapter 5.3.4 presents requirements for gNB setup and configuration. Some key statements include:

- gNB setup and configuration relying on operations and maintenance (O&A) system needs to be authenticated and authorized by gNB. This prevents external parties from modifying the gNB settings by using local or remote access methods. Furthermore, the communications between gNB and O&A must protect confidentiality, integrity, and relay.
- gNB needs to support certificate mechanism detailed in 3GPP TS 33.310. Nevertheless, MNO decides the strategy for the utilization of the certificates.

- More details of the security measures of gNB are summarized in ETSI 133 501, and the security associations between gNB, O&A, and 5G core network are detailed in 3GPP TS 33.210 and TS 33.310.

10.6   5G Network Interfaces and Elements

This section summarizes the 5G-specific network reference points between the network functions. For the complete list of the reference points, please refer to Chapter 6, "Core Network."

10.6.1   Interfaces

Figure 10.10 summarizes the 5G interfaces in a reference point format as interpreted for the 3GPP TR 23.799.

Please refer to 3GPP TR 23.799 and 3GPP S2-167226 for more detailed information on the nonroaming and roaming scenarios, including local breakout and home routed cases, concurrent access to local and central data networks with multiple packet data unit (PDU) session as well as concurrent access to two data networks in a single PDU session scenario.

10.6.2   5G Functional Elements

There are various new elements in the 5G system. Some of the 5G element, i.e. networks functions, have completely new roles while others have similar, enhanced functions compared to previous network generations. Table 10.2 summarizes the 5G elements.

Figure 10.10 The 5G reference points.

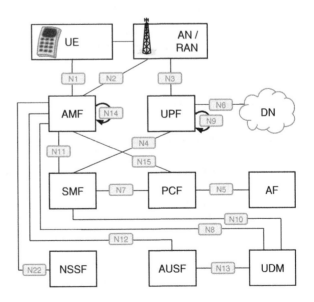

Table 10.2 The 5G network deployment's functional elements.

Entity	Description
AF	Application function requests dynamic policies and/or charging control.
AMF	Access and mobility management function supports (i) termination of RAN CP interface (*NG2*); (ii) termination of NAS (NG1), NAS ciphering and integrity protection; (iii) mobility management; (iv) Lawful Interception (LI); (v) transparent proxy for routing access authentication and SM messages; (vi) access authentication; (vii) access authorization; (viii) Security Anchor Function (SEA). AMF interacts with UDM and UE and receives intermediate key which is a result of the UE authentication process. If USIM based authentication is applied, AMF gets the security material from the UDM; (ix) Security Context Management (SCM). The SEA provides a key to SCM for the derivation of access-network specific keys.
AUSF	Authentication server function performs authentication processes with the UE.
DN	Data network can be e.g. Internet, operator services or third-party services.
PCF	Policy control function supports (i) unified policy framework to govern network behavior; (ii) policy rules to control plane functions that enforce them.
RAN	Radio access network.
SMF	Session Management Control Function supports (i) session management; (ii) UE IP address allocation and management with an optional authorization; (iii) selection and control of UP function; (iv) termination of interfaces toward policy control and charging function; (v) control-part of policy enforcement and QoS; (vi) LI; (vii) termination of SM parts of NAS messages; (viii) downlink data notification; (ix) initiator of AN-specific SM information, sent via AMF over *NG2* to AN; (x) roaming functionality; (xi) handle local enforcement to apply QoS SLAs (VPLMN), charging data collection and related interfacing, and LI.
UDM	Unified data management supports (i) Authentication Credential Repository and Processing Function (ARPF), which stores long-term security credentials that are the base for AKA authentication; (ii) storing of subscription information.
UE	User equipment.
UPF	User plane function supports (i) anchor point for intra-/inter-RAT mobility (when applicable); (ii) external PDU session point of interconnect (e.g. IP); (iii) packet routing and forwarding; (iv) QoS handling for user plane; (v) packet inspection and policy rule enforcement; (vi) LI (UP collection); (vii) traffic accounting and reporting; (viii) interaction with external DN for transport of signaling for PDU session authorization/authentication by external DN.

10.7 Core Deployment

The 5G technical standards allow gradual deployment of the new network infrastructure. The NSA option has been defined on December 2017 providing the operators with the possibility to deploy 5G earlier in the market compared to the IMT-2020-compliant networks as the respective requirements will be ready around 2020.

The NSA options specify only New Radio, which will work in tandem with 4G radio, relying completely on the 4G core. This scenario facilitates the 5G hotspots while the 4G provides the overlay network coverage. This option is known as E-UTRAN-NR dual connectivity (EN-DC).

The next step of the 3GPP specifications has been the completion of Release 15, also referred to as 5G Phase 1, by adding the specifications for the complete 5G radio (including standalone radio functionality), 5G core, the completely renewed security architecture and functionality, and respective supporting specifications. Phase 1 provides the eMBB mode.

The evolution of 5G includes the additions of new 5G functionalities in Release 16, also referred to as 5G Phase 2. It completes the 5G with the aim to comply with the ITU-R IMT-2020 requirements, including specifications for the mIoT (or, massive machine-type communication, mMTC) and ultra-reliable low-latency communications (URLLCs). The aimed freezing schedule of Release 16 is set for December 2019, which coincides with the ITU time schedule for the evaluation of the 5G candidate technologies during 2020.

10.7.1 Initial Phase: Reusing Existing Infrastructure

Cost-efficient 5G deployment relies on the existing 4G infrastructure. The 3GPP specifications facilitate this gradual transition via multiple options summarized in previous sections. The interoperability and interworking between 4G and 5G is thus an elemental part of the 5G deployment scenarios.

The 5G radio network can be connected to 4G or 5G core. Furthermore, the 4G radio network can also be utilized as a part of the transition. The 5G network is referred to standalone RAN when it consists solely of respective 5G base stations, gNB elements. The non-standalone options refer to the combination of 4G and 5G base stations.

Option 3 is one of the most practical from the scenarios designed to facilitate the non-standalone deployments. This can also be referred to as EN-DC architecture. IT is especially optimal for deployments prior to the IMT-2020-capable networks that will be reality as of 2020.

The EN-DC utilizes both LTE radio network and EPC core network of 4G. This means that the 5G-type of services will be available without replacing the network infrastructure. In the EN-DC architecture, the LTE eNB infrastructure acts as a master RAN, and the 5G NR infrastructure can be added gradually in such a way that the UE is connected via dual connectivity to LTE and NR.

This scenario requires minor modifications in 4G EPS based on Release 15 technical specifications. These modifications include the separation of the user and signaling planes for the SGW and PGW elements, and a new method to handle the quality of service (QoS) class identifier (QCI), which is defined over the whole system to ensure the compliance with the low latency values below 10 ms. As a result, the 4G EPC can support 5G use cases in mIoT, eMBB, and URLLC categories.

Figure 10.11 presents an example of the non-standalone EN-DC architecture based on E-UTRA and NR infrastructure [6]. Furthermore, continuing Ref. [6], the following three figures give examples of the deployment of NR and E-UTRA (NE-DC) scenario according to Option 4, NG-RAN and EN-DC scenario (NG-EN-DC) according to Option 7, and Enhanced Packet System (EPS) and 5GS interworking (Figures 10.12–10.14).

10.8 CoMP

As stated in [7], the idea of enhanced signaling for Inter-eNB CoMP is to provide a way to coordinate with the base station to minimize their interference. Inter-eNB CoMP

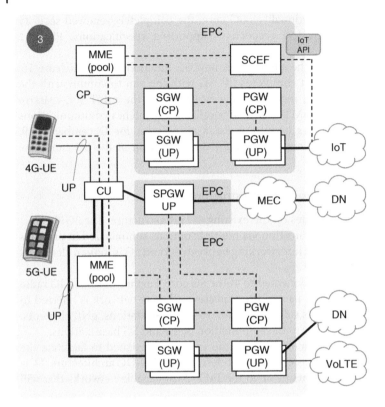

Figure 10.11 Dual connectivity for non-standalone scenario of E-UTRA and New Radio (EN-DC).

may thus provide enhanced performance via augmented radio coverage and respective data services within the cell and in cell-edge areas. The CoMP has been researched and summarized in 3GPP TR TR.36.819, in phase Release 11. This resulted in Release 12 solution, which is a centralized master-slave option of the CoMP. In this solution, a slave eNB within a cluster facilitates coordination data for a Centralized Coordination Function (CCF). The centralized approach requires a new node and interface, which is challenging in practice. Thus, the alternative option has been basing on a distributed peer-to-peer solution, which refers to a scenario where each eNB is able to exchange coordination information with the neighbors of a cluster via already existing *X2* interface in a decentralized manner.

Figure 10.15 summarizes the CoMP modes, which are relevant also in 5G era to optimize the radio performance.

10.9 Measurements

10.9.1 Overall View

Network performance and functional measurements belong to the essential task set in the 5G network planning, deployment and optimization. There are commercial devices

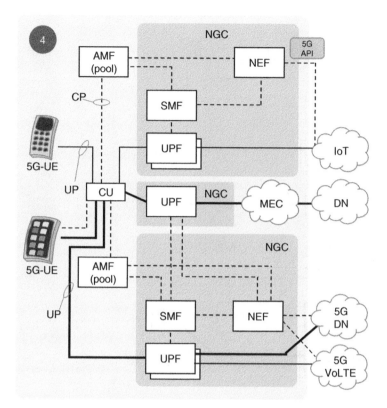

Figure 10.12 NR and E-UTRA (NE-DC) scenario referred to as Option 4.

for testing and analysis of the measurement data. In fact, even the first 5G standards were not ready yet before the freezing of the 3GPP Release 15, the pre-standard-based test equipment have played a key role in the proofing of different concepts, which has been a base for the actual selection of the most adequate technical solutions into the specifications. Test equipment is typically based on sufficiently performant hardware and a set of software components, which can be upgraded as the standards evolve.

5G is based on the already existing frequency bands as well as completely new, higher frequencies. The new bands include 6, 28, and 38 GHz. ITU-R coordinates the efforts to agree the global policies for the frequencies via the WRC (World Radio Conference), from which the essential event for the selection of the extended bands takes place at WRC-19.

In the initial phase of the 5G, the bands on 28 and 38 GHz are estimated to be popular in the United States, whereas 26 GHz is used in Europe and China.

Besides the high frequencies in mm-Wave area, many other candidates are seen possibly on 600–900 MHz, 1.5, 2.1, 2.3, and 2.6 GHz. These bands may be especially useful in applications requiring less capacity, such as IoT, as they provide larger coverage areas.

Among other stakeholders, the GSA has reasoned that due to the radiowave propagation characteristics, the bands 3300–4200 MHz and 4400–4990 MHz are especially suitable for the initial phase of the 5G. They facilitate the balance between the capacity and coverage areas by maintaining the expedited deployment schedules of the 5G.

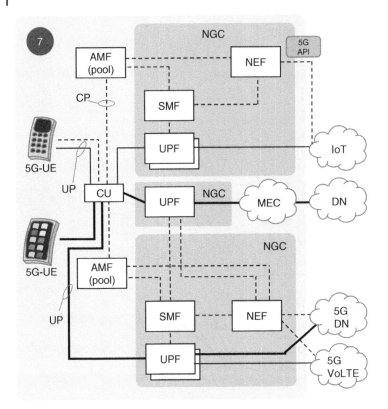

Figure 10.13 NG RAN and EN-DC scenario (NG-EN-DC) referred to as Option 7.

The prestandard-based 5G networks are expected to be seen during 2018, and the non-standalone mode of the 3GPP Release 15, which was completed in December 2017, has resulted in the possibility to expedite the first 5G deployments during 2019. As for the standalone networks, Japan, for example, has a goal of providing the ITU-R's IMT-2020 compliant 5G services as of 2020, the aim being to have commercial services available by the summer Olympic games.

It can be assumed that there will be a multitude of existing and new frequency bands in 5G deployments. At the same time, the MNOs evaluate the feasibility to operate 5G networks in a parallel fashion with previous generations, the latter complementing the services of the legacy users. At some point, though, the old technology may convert obsolete as can be seen from the decommissioning of GSM network of AT&T in the United States.

The correct moment to adjust the capacity between the networks is important as the respective optimization has direct and indirect impacts on the business models. The newer technologies are more spectral efficient while the older ones may still support a significant customer base relying on legacy devices.

The first-phase 5G networks are based on sub-6 GHz radio bands. They facilitate fast and fluent deployment and service coverage areas while the higher frequency bands provide with capacity extension via small cell deployments.

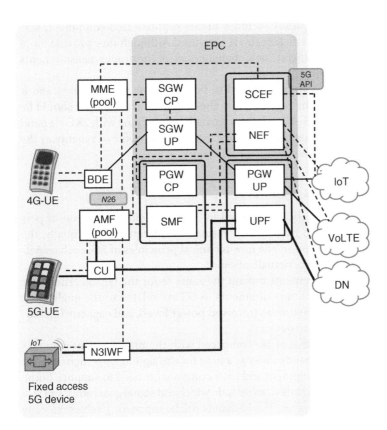

Figure 10.14 Deployment for EPS and 5GS interworking.

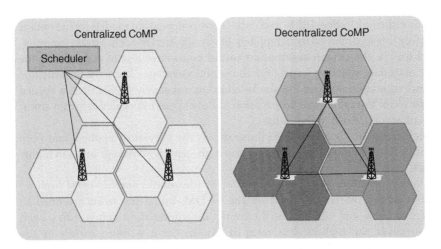

Figure 10.15 The CoMP options as defined in 3GPP Release 12 are still relevant in 5G.

The longer-run 5G strategy typically includes highly sophisticated equipment, services, and methods such as more efficient smart antennas. Although they provide more performant networks, they also generate new challenges as for the proper measurements of their functioning and performance gains.

In Europe, the 26 GHz band can be assumed to be one of the popular ones, and it will probably be harmonized in the region before the WRC 19 takes place. It should be noted that the 26 GHz band (24.25–27.5 GHz) is partially overlapping with 28 GHz band (26.5–29.5 GHz), which means that they can have significant benefits of synergy as the equipment manufacturers can provide compatible bands.

10.9.2 Measurement Principles of 5G

The new, forthcoming frequency bands adopted in 5G will require new level of performance and functionalities also from the RF measurement equipment. Equally, the equipment must consider and support the new technical principles of 5G, such as network slicing and network functions virtualization.

The basics of the radio measurements remain the same as for the typical criteria for the results; some of the still useful measurements in 5G are related to the quality (bit rates, bit error rates), radio coverage areas (received power level), and capacity (including simultaneously communication devices).

One of the remarkable differences of 5G compared with the previous generations is the significantly increased radio bandwidths as a base for the significantly increased bit rates, which the measurement equipment and their components need to support. Some examples of filed and laboratory test devices include wideband signal generators and signal analyzers, including the mm-waves. The 5G bands will be supporting values typically from 500 MHz up to 2 GHz. The actual novelty 5G bands are between 6 and 60 GHz, some options being even higher. The same principle applies in 5G as has been in all the previous generations; the higher the frequency, the smaller the coverage area. Thus, the frequencies over 6 GHz typically apply for small cell environment, in many cases limited to a building's floor or a single room.

The feasible radio link budget (RLB) and radio-wave propagation models include many already existing ones, combined by completely newly adjusted models. For the frequencies beyond 6 GHz, the default assumption would be to have line of sight (LOS) the obstructing materials affecting largely for the useful coverage. The materials include buildings and vegetation such as forests. In addition, especially 60 GHz has a special characteristic as for the radiowave propagation as the oxygen molecules cause an attenuation peak of around 20 dB in that range.

For the high-frequency bands, the 5G thus extends the previously applied femto cell concept in the efforts to form high-data rate and high-capacity coverage areas in highly limited areas.

The previous generation, 4G, is based on OFDM (downlink) and SC-FDM (uplink) modulation schemes. 5G is still based on the OFDM, but it has been extended to cover both downlink and uplink. Furthermore, the modulation scheme has been optimized to cover the highest data rates of 5G, so there is more variety of the subcarrier spacing and higher-grade modulation, up to 256-QAM. This outcome is a balance of the performance, device complexity, and other aspects. Some of the evaluated modulation schemes that led into this conclusion of the optimal balance of OFDM were the 4G SC-FDM (Single-Carrier Frequency Division Multiplex) and OFDM based on cyclic prefix, and the Unique-Word DFT-Spread-OFDM, GFDM, UW-OFDM, Resource-Block-Filtered OFDM, Universal Filter Multi-Carrier (UFMC) and Filter-Bank-Multi-carrier (FBMC).

In the practical measurements of the performance of the modulation (including the previous efforts to evaluate the candidate technologies), some of the useful measures are related to the energy efficiency indicated by peak-to-average power ratio (PAPR), spectral efficiency indicated by the out of band leakage (OOB), and the reliability of the link indicated by bit error rate (BER) applied in the multi-path propagation environment. There are various needs for 5G measurement equipment, including multiuser support, MIMO performance (multiple-in, multiple-out antenna arrays), transmission latency, and quality of the synchronization, to mention some.

The 5G system is dynamic and flexible. As a benefit of the network slicing and network functions virtualization, a single network can thus provide a variety of different levels of performance levels for the users, including the same user, in terms of QoS, protection level, and performance. The OFDM is also especially suitable for supporting such scalable approach from the radio interface, with flexible subcarriers.

Summarizing information from Refs. [8–10], Table 10.3 lists some measurement types in 5G.

The laboratory and pilot networks have provided valuable information on the expected performance of the 5G prior to the actual deployment of the networks. Part of this evolution is related to the development of the measurement equipment, which can be utilized for the investigation of the proof of concepts, and later, for the evaluation of the performance of the radio and transmission networks. Sufficiently reliable and performant set of test equipment plays an essential role in both pre-deployments as well as in the actual deployment phases to obtain realistic results.

10.9.3 NGPaaS

Apart from the actual 5G network architectures and their deployment, based eventually on the native, standalone option and service-based architecture, there are accompanying concepts that support the 5G strategies. One of these concepts is the Next Generation Platform-as-a-Service (NGPaaS). Its aim is to offer personalized and flexible platform for the further development of 5G, which also ensures better interoperability of the stakeholders.

Nokia Bell Labs is in a leading position to drive this development, which is a coalition of industry and academia and which is at the same time part of the 5G-PPP (5G Infrastructure Public-Private Partnership). The latter is aimed to get closer the European Union and telecom companies.

The PaaS concept is designed to be more compatible with cloud ecosystems compared to the previous Infrastructure as a Service (IaaS) model . It is thus expected to be suitable for the needs of 5G related to the Virtual Network Function (VNF). VNF, in turn, needs to support the demanding requirements of performance of 5G in terms of capacity, latency, and reliability. At present, the solutions do not support such requirements in an optimal way by relying on cloud, so the aim of NGPaaS is to comply with the criteria set up by ITU-R IMT-2020.

Table 10.3 Examples of 5G measurement equipment.

Type	Examples and suitability
Signal analyzer	Signal and spectral analyzer are designed to support frequency ranges, which may be by default several tens of GHz. In many cases, the range can be further extended by down-conversion. The demodulation of bandwidth may be several hundreds of GHz in the basic models, while the range can be extended up to the maximum of 2 GHz for signal analysis by native HW/SW or by relying on e.g. external oscilloscope. The equipment may analyze digitally modulated singe subcarriers in bit-level or limited to generic OFDM signal analysis. The supported frequency bands may vary from basic models' sub-6 GHz up to highest foreseen bands in the 100 GHz range.
Signal generator	The vector signal generators can be applied to the simulation of wideband signals. The equipment may have native 2 GHz modulation bandwidth and the supported band set may include a variety of options. The equipment may also be able to generate variety of waveforms, apart from the selected 5G OFDM-based ones, such as BMC, UFMC, GFDM, and f-OFDM. The equipment may also support the generation of intereffects and multiple component carriers for carrier-aggregated simulations, and modulation schemes up to the maximum defined 5G variant of 256-QAM. Some measurement items of these equipment include channel power level, co-channel power, and error vector magnitude.
OTA measurements	OTA (over-the-air) test chamber is typically fixed but can also be a mobile variant. It is meant for the measurements of 5G antennas. The related test system may include means to measure both passive and active antennas. Some measurements are the RF patterns of complex antennas in 3D domain. The system may also support the measurements of the performance of radio transceivers. Typically, these types of measurements support all the defined radio frequencies up to the maximum, around 87 GHz. The measurements indicate the effects of 5G RF OTA parameters.
Network emulator	The network emulator can be useful for the performance measurements of multiple subchannels of the OFDM, combined with vector signal generator, as well as signal and spectrum analyzer.

References

1 Verizon, "Verizon 5G Technical Forum," Verizon, 2018. http://www.5gtf.net. [Accessed 26 July 2018].

2 DoCoMo, "5G Trial Site," DoCoMo, 2018. www.nttdocomo.co.jp/english/corporate/technology/rd/docomo5g/trial_site/index.html. [Accessed 26 July 2018].

3 Telia Company, "First live 5G use cases deployed today," Telia Company, 2018. http://www.teliacompany.com/en/news/news-articles/2017/first-live-5g-use-cases-deployed-today. [Accessed 26 July 2018].

4 European Union, "The 5G Infrastructure Public Private Partnership," European Union, 2018. https://5g-ppp.eu/5g-trials-roadmap. [Accessed 26 July 2018].

5 Qualcomm, "Making 5G NR a reality," Qualcomm, San Diego, USA, 2016-12.

6 NEC, "Making 5G a relaity," NEC, Tokyo, 2018.

7 5G Americas, "Wireless technology evolution towards 5G: 3GPP Release 13 to Release 15 and beyond," 5G Americas, 2017.

8 National Instruments, "NI Demonstrates Wideband 5G Waveform Generation and Measurement Technology for 5G Test Applications," National Instruments, 5 06 2017. http://www.ni.com/newsroom/release/ni-demonstrates-wideband-5g-waveform-generation-and-measurement-technology-for-5g-test-applications/en. [Accessed 26 July 2018].

9 Sarokal, "Sarokal Test Systems news," Sarokal, 2018. http://www.sarokal.fi/sarokal+test+systems+to+participate+in+building+finland-s+first+5g+test+network. [Accessed 26 July 2018].

10 Rohde & Schwartz, "5G Wideband Signal Generation and Signal Analysis," Rohde & Schwartz, 2018. https://www.rohde-schwarz.com/us/solutions/wireless-communications/5g/5g-wideband-signals/5g-wideband-signals_229599.html. [Accessed 26 July 2018].

Index

Page numbers referring to figures are in *italics* and those referring to tables in **bold**.

5G Explained: Security and Deployment of Advanced Mobile Communications, First Edition.
Jyrki T. J. Penttinen.
© 2019 John Wiley & Sons Ltd. Published 2019 by John Wiley & Sons Ltd.

Printed and bound by CPI Group (UK) Ltd, Croydon, CR0 4YY

16/04/2025

14658391-0004